HUMAN ANATOMY
COLORING BOOK FOR KIDS

Copyright ©
All right reserved

THIS BOOK BELONGS TO:

_____

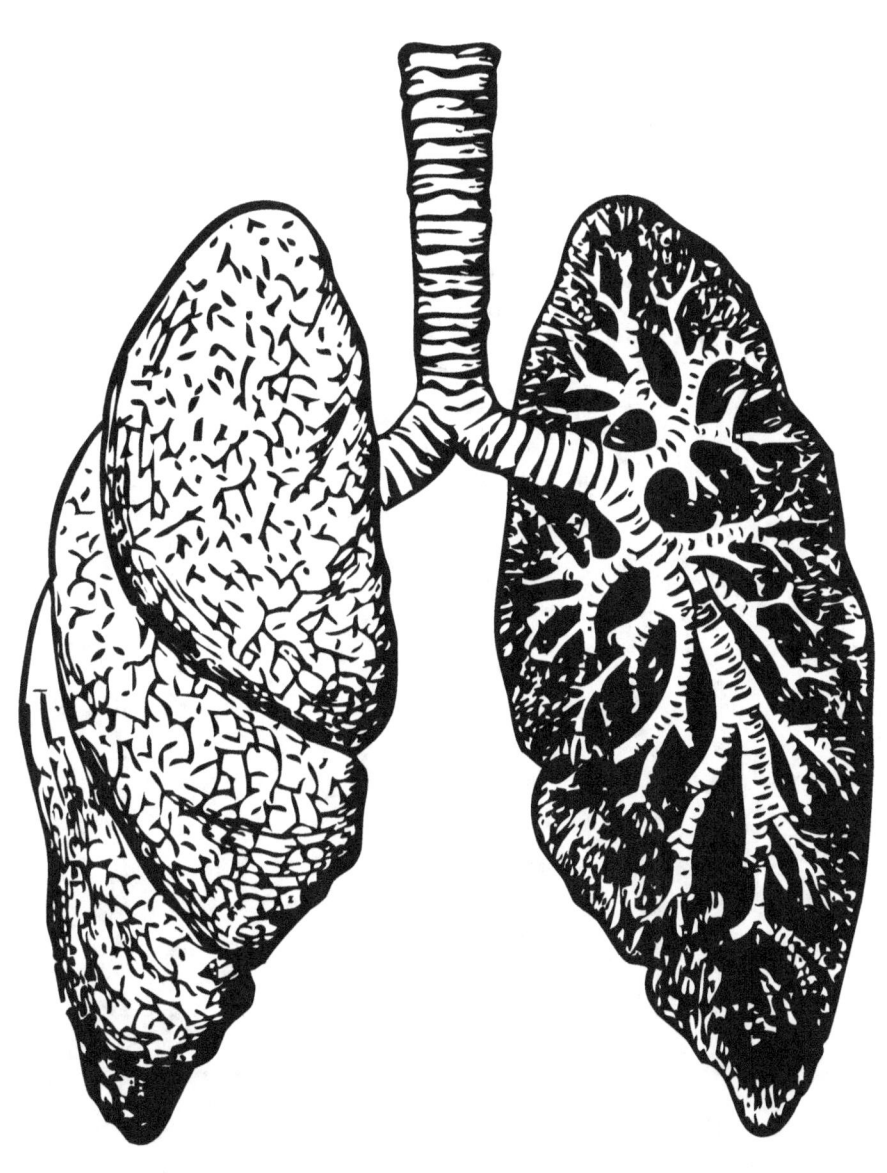

TEST YOUR COLORS

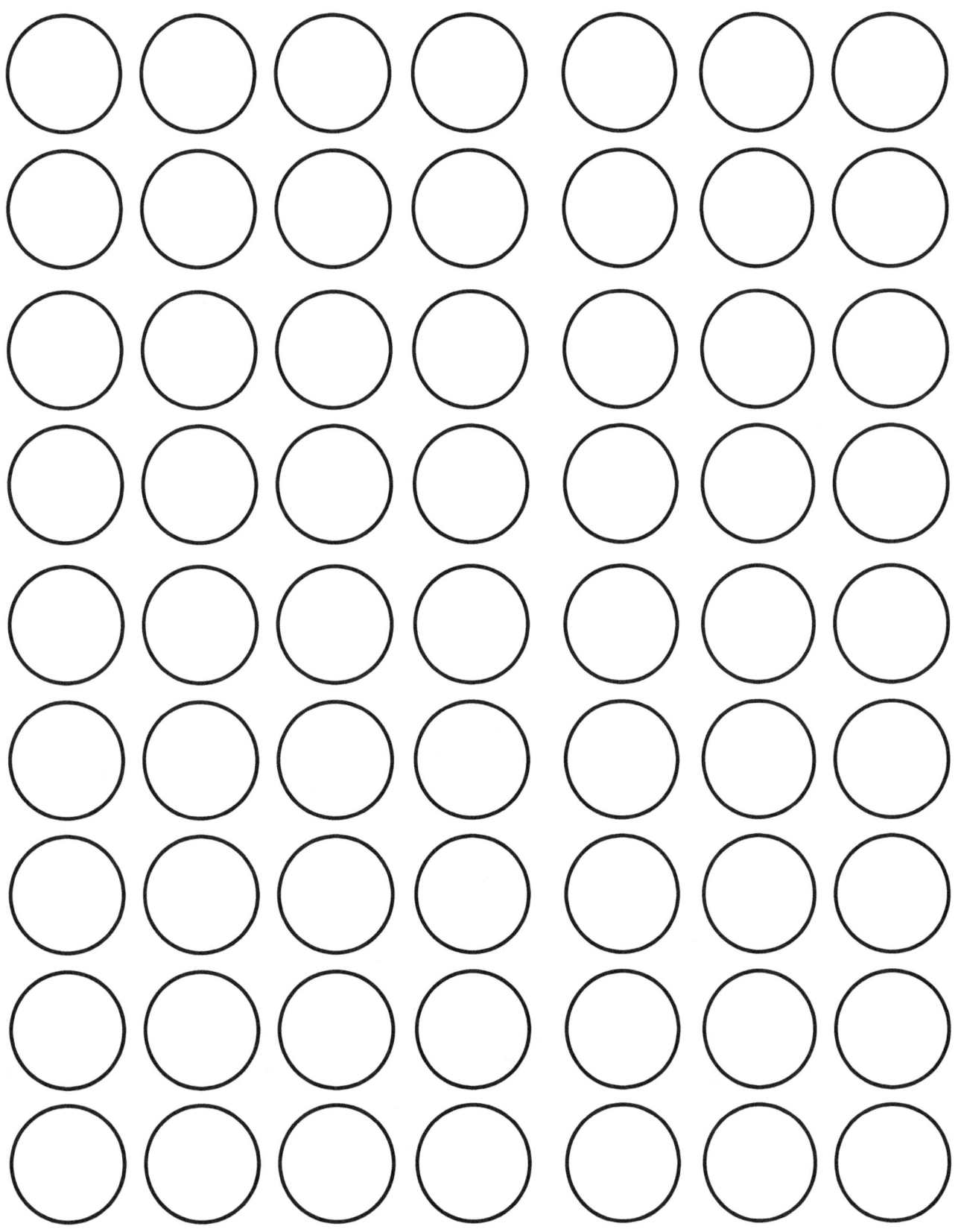

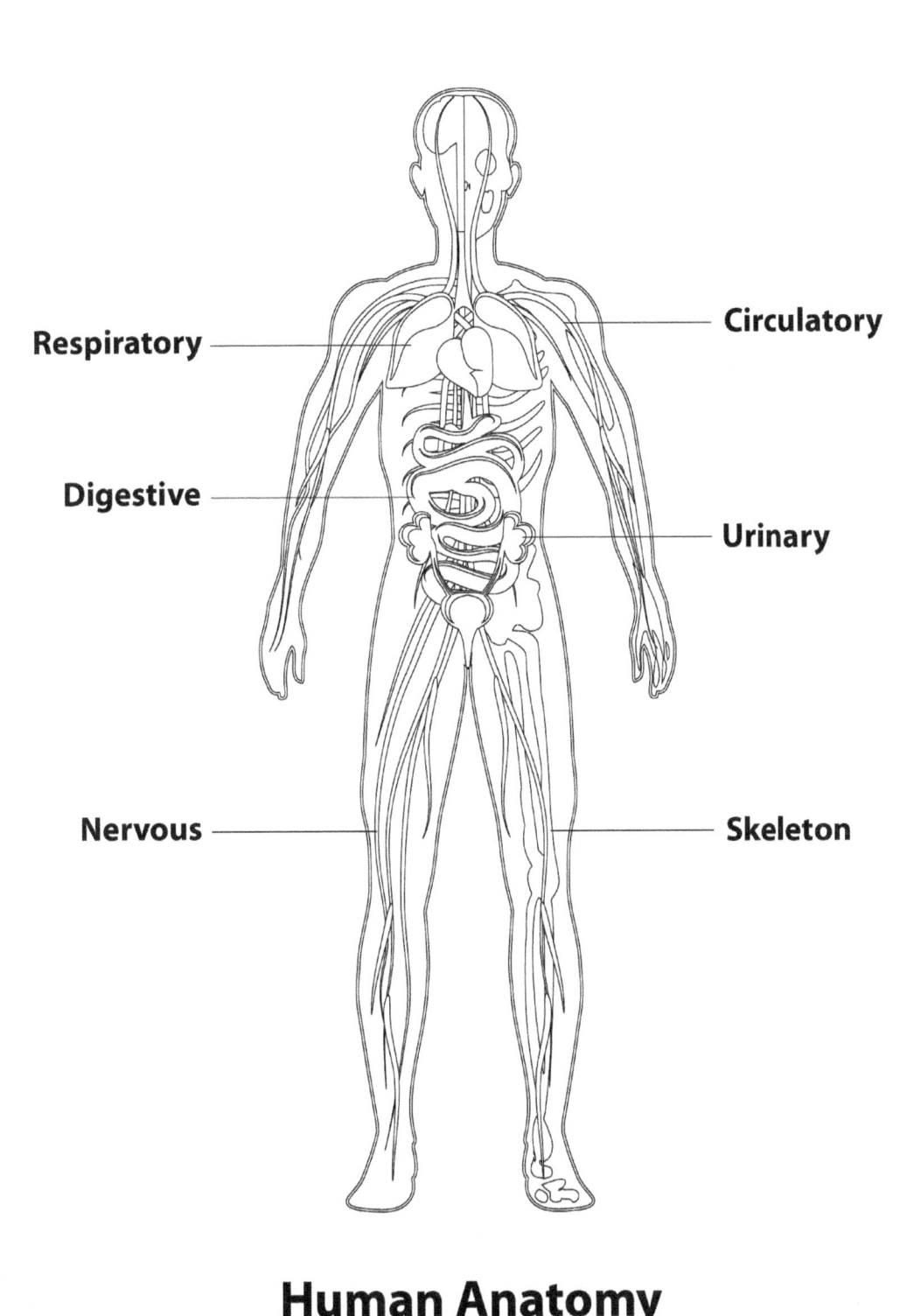

Human Anatomy

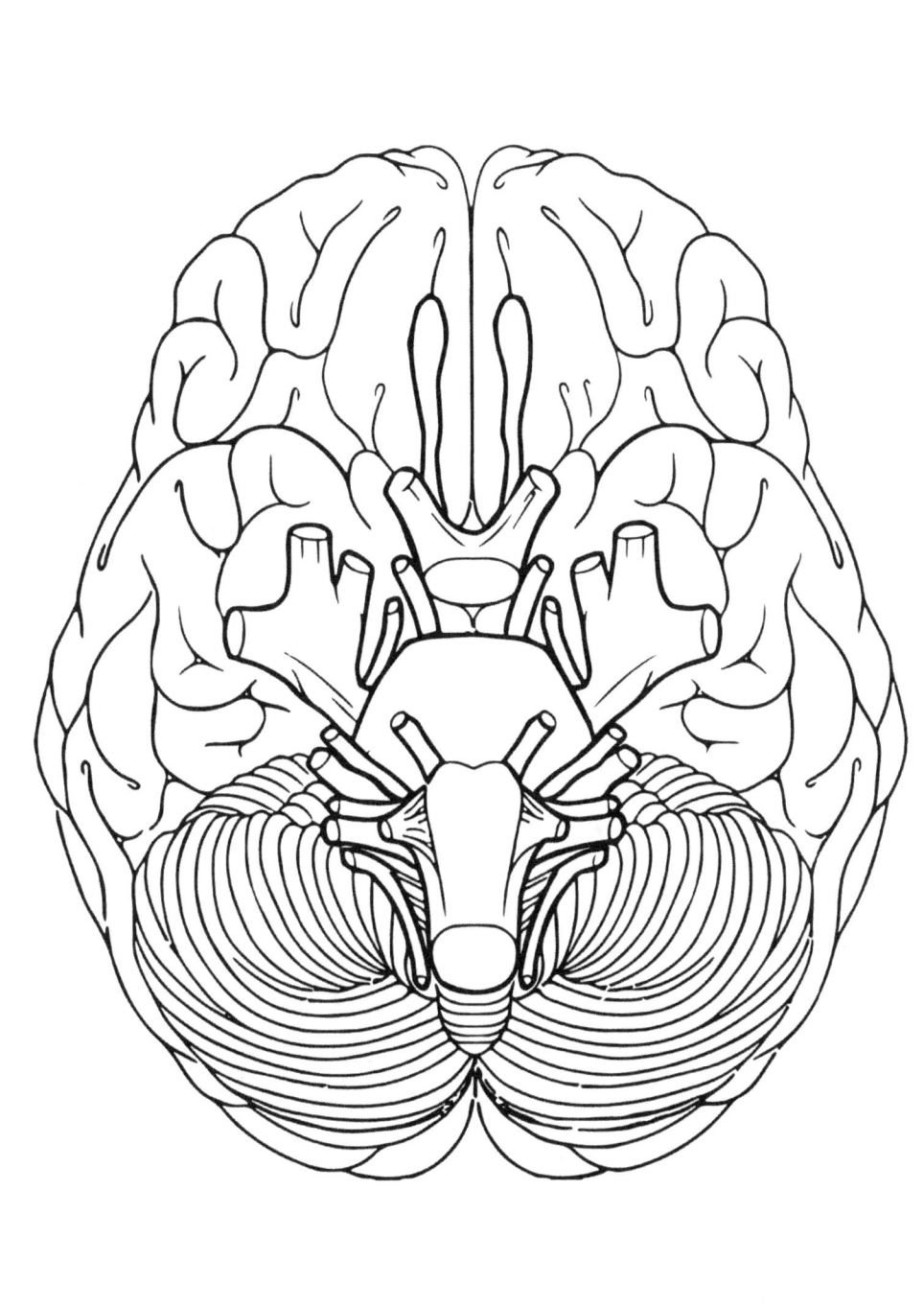

Brain

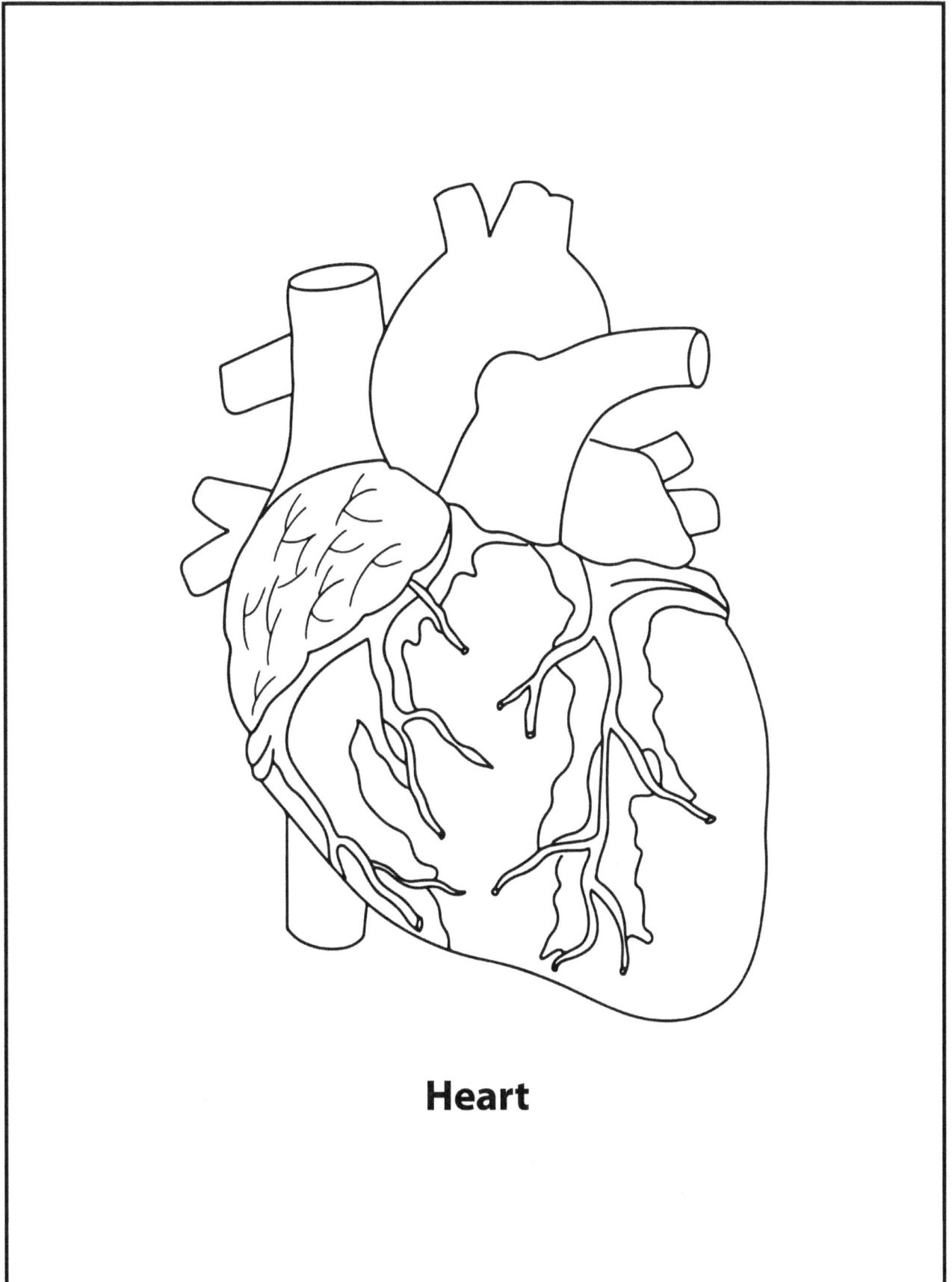

Heart

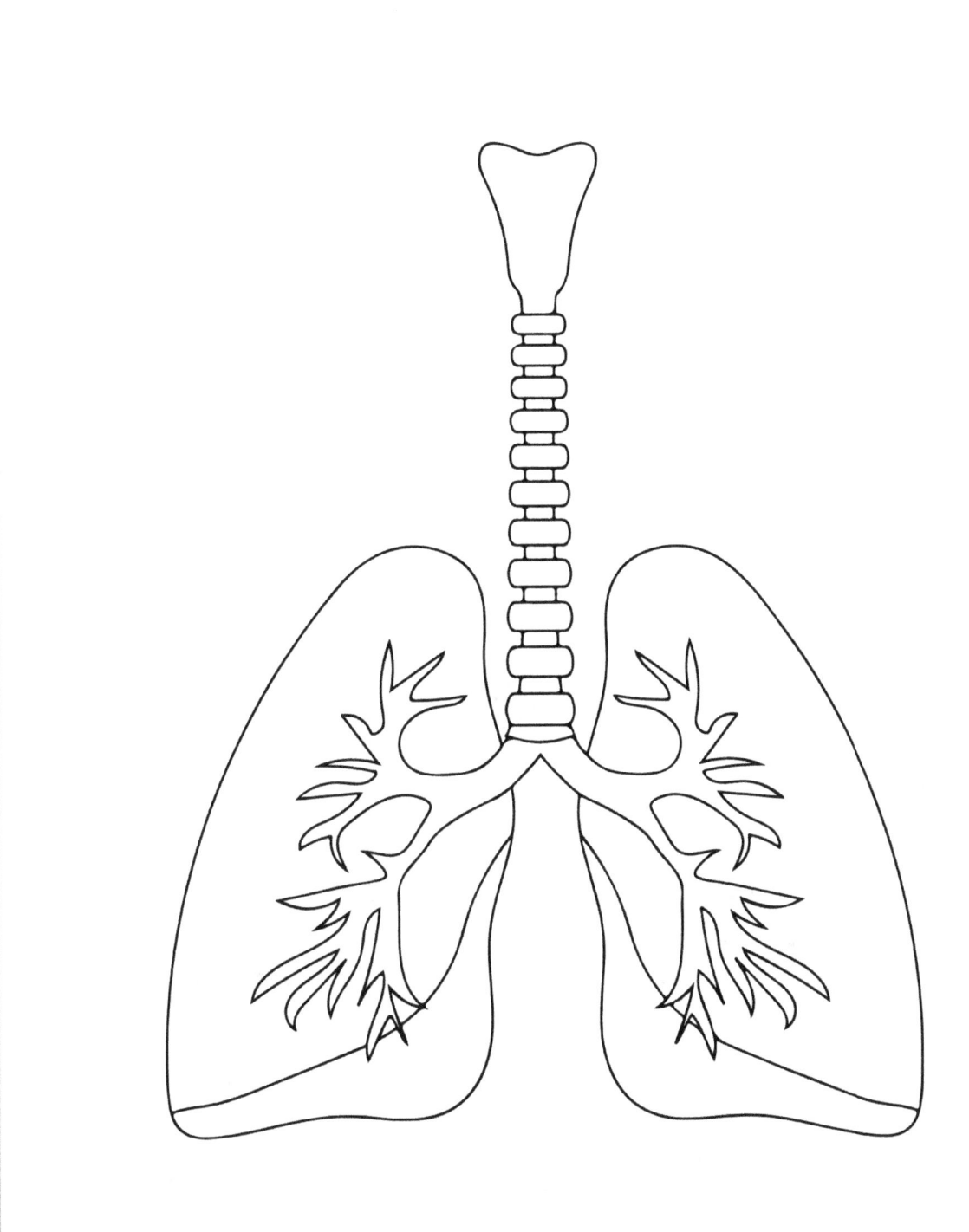

Lungs

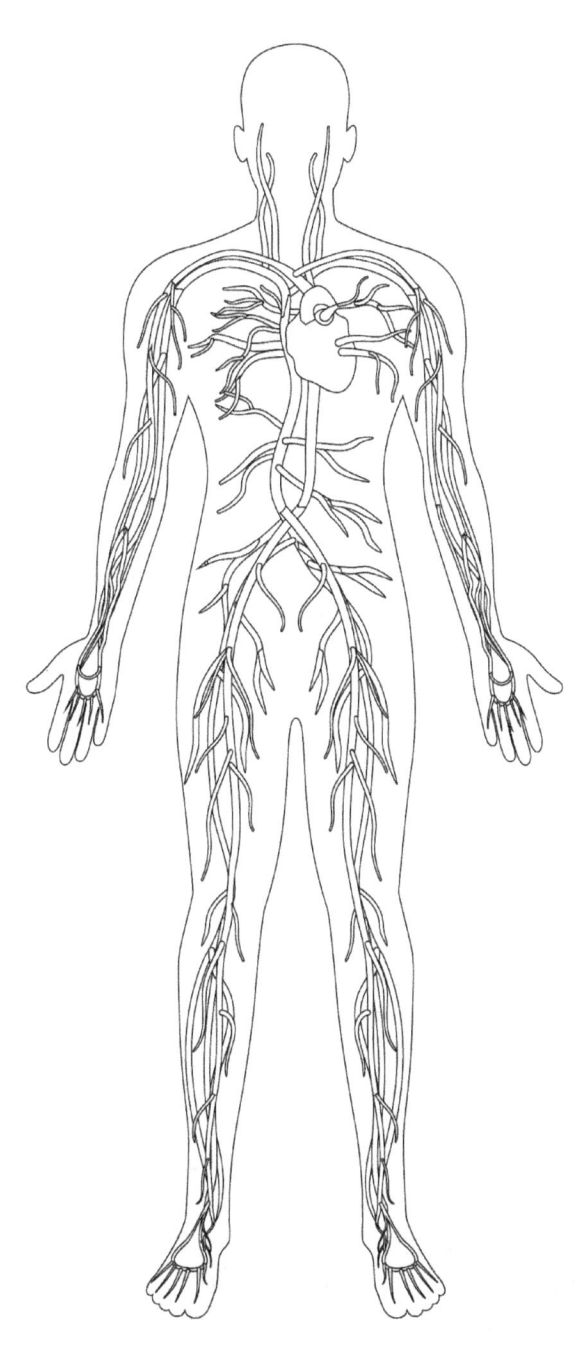

Circulatory System

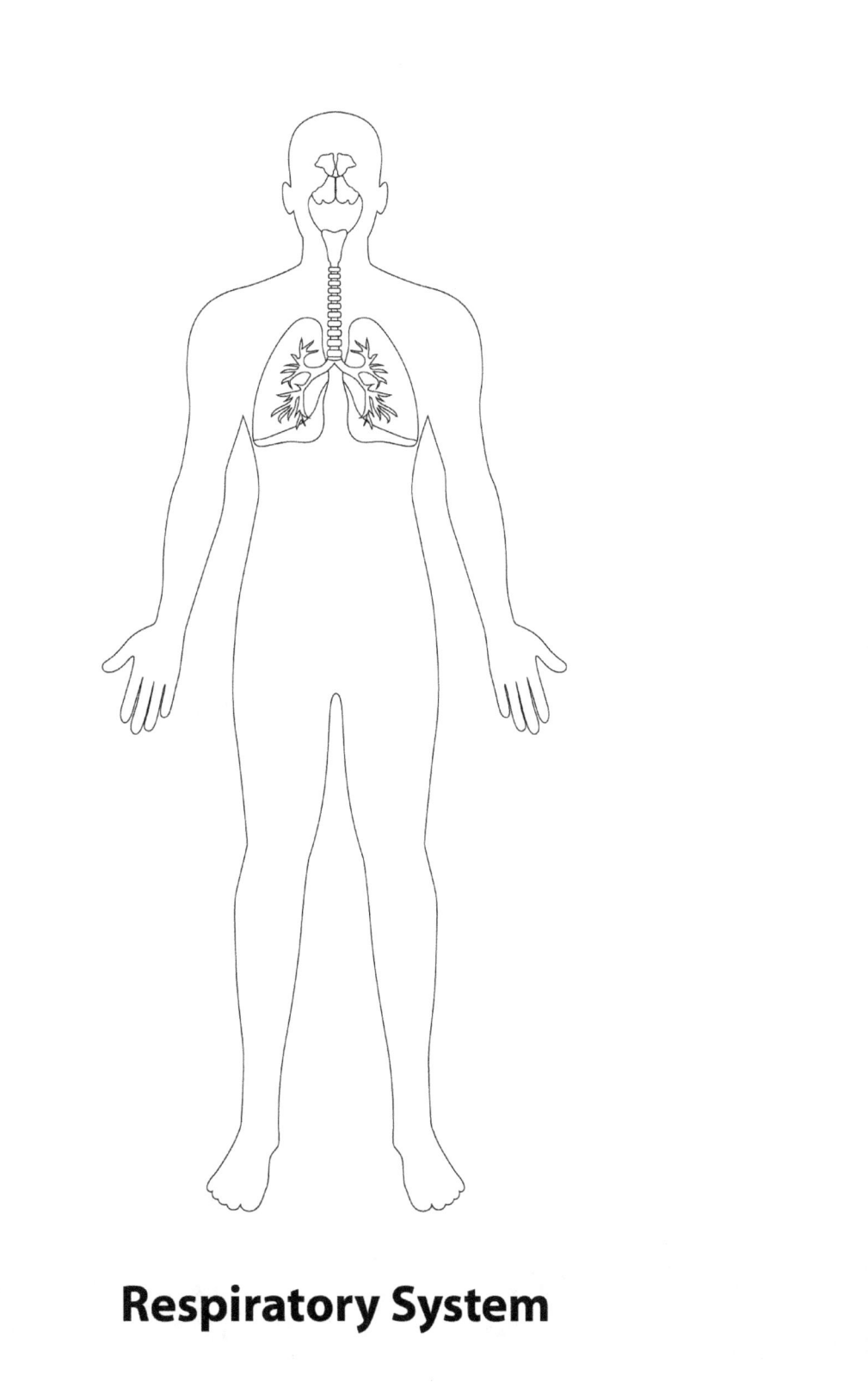

Respiratory System

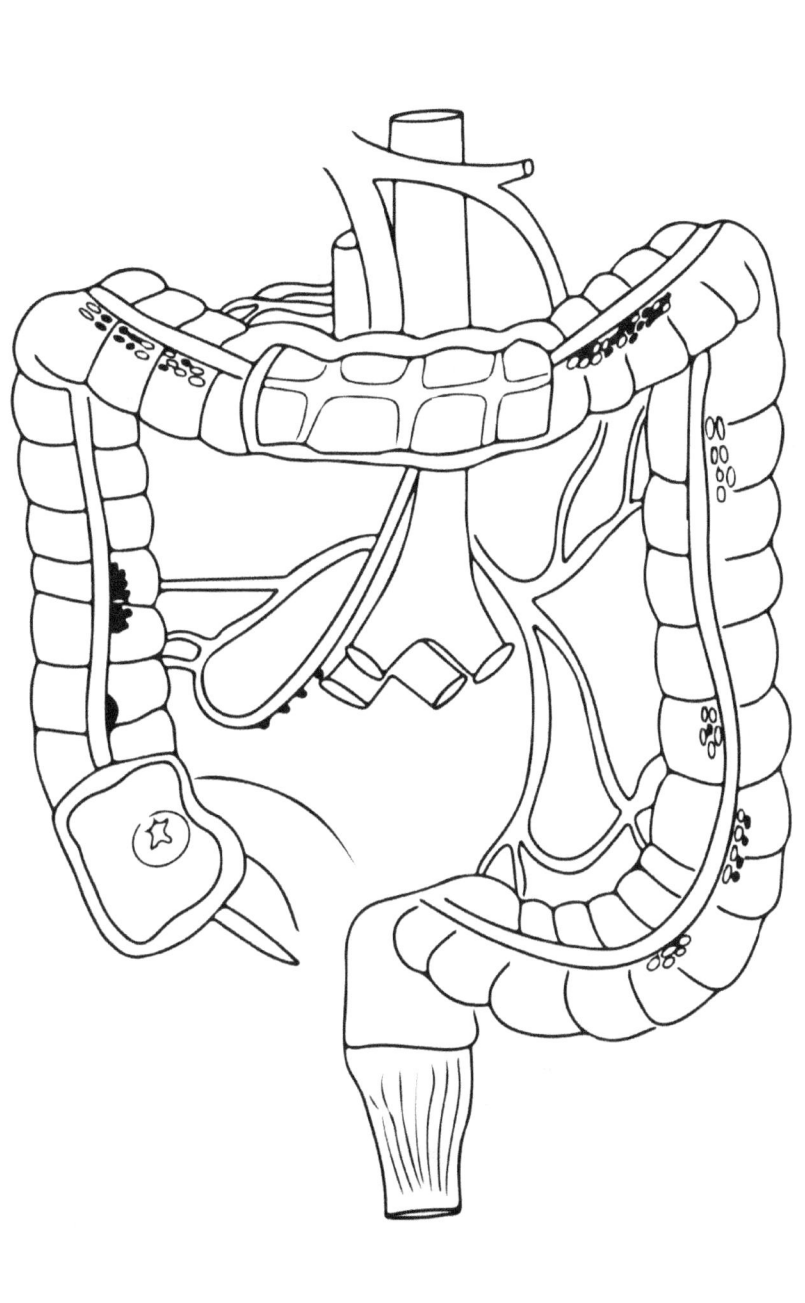

Intestine

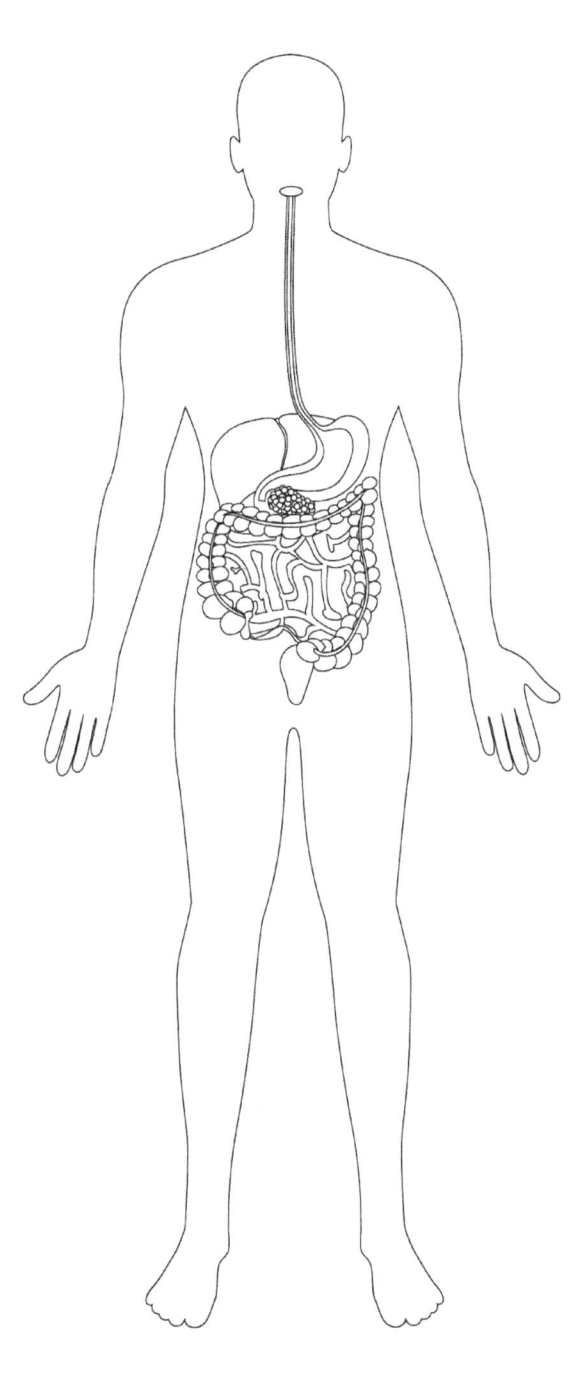

Digestive System

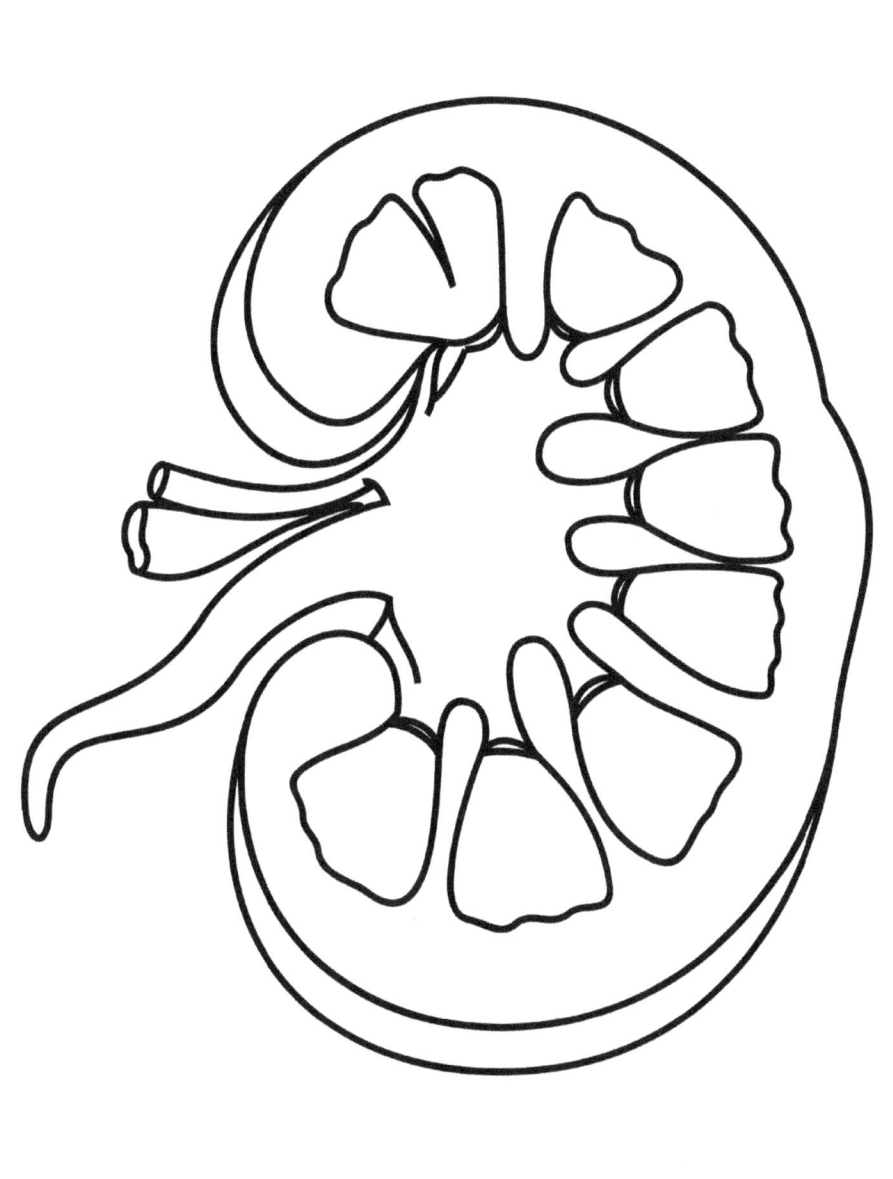

Kidney

Nerve Cell

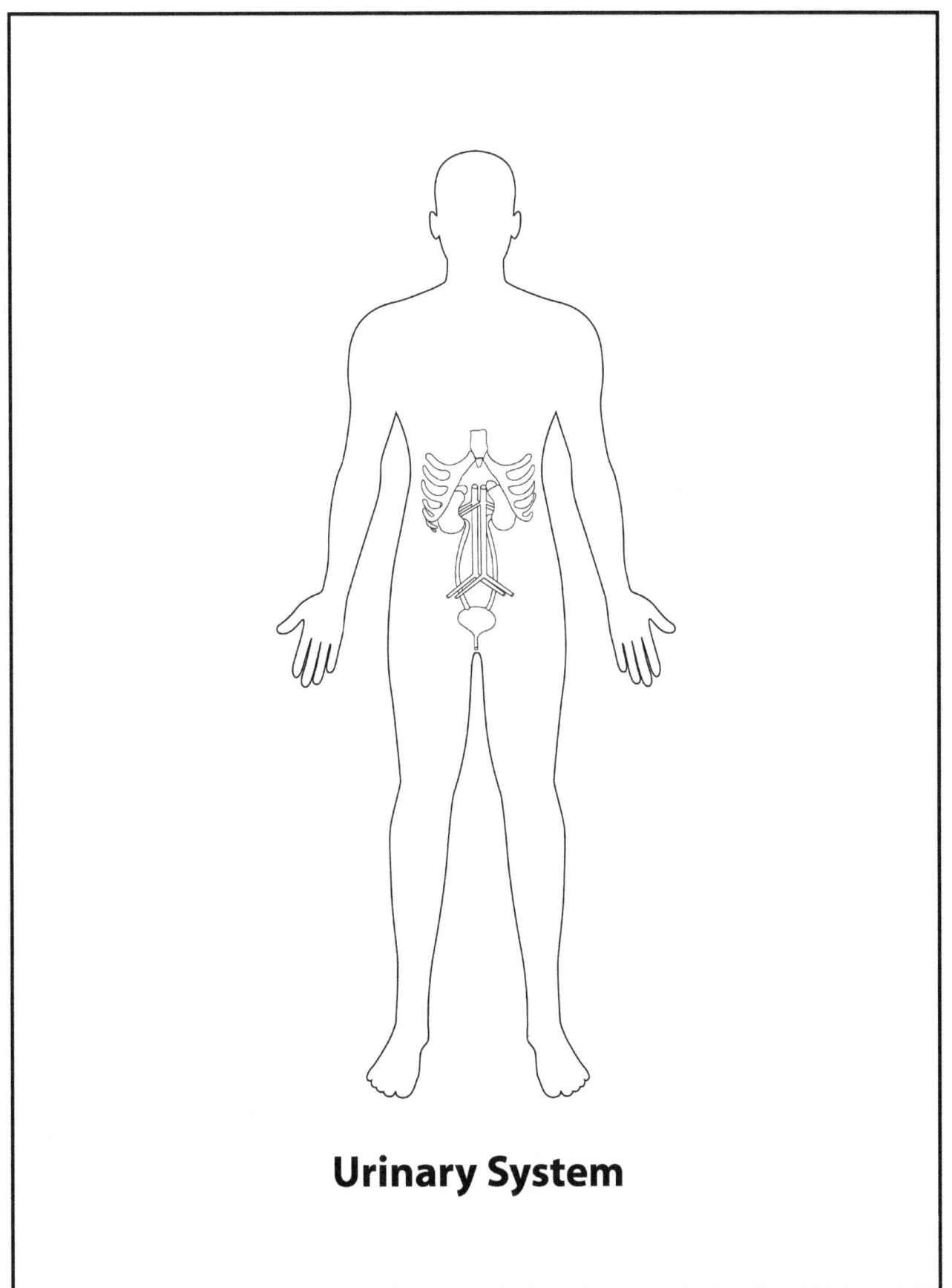

Urinary System

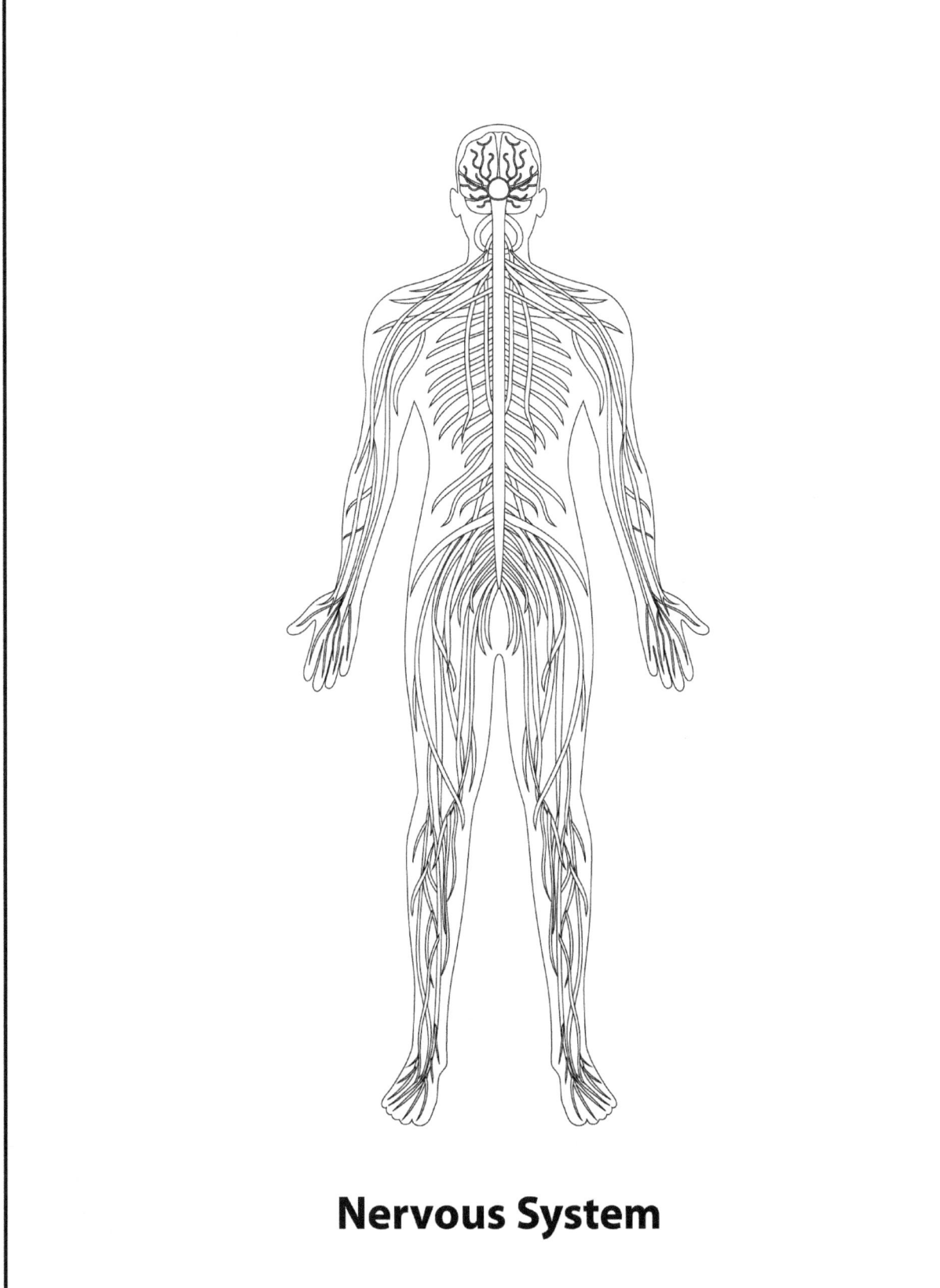

Nervous System

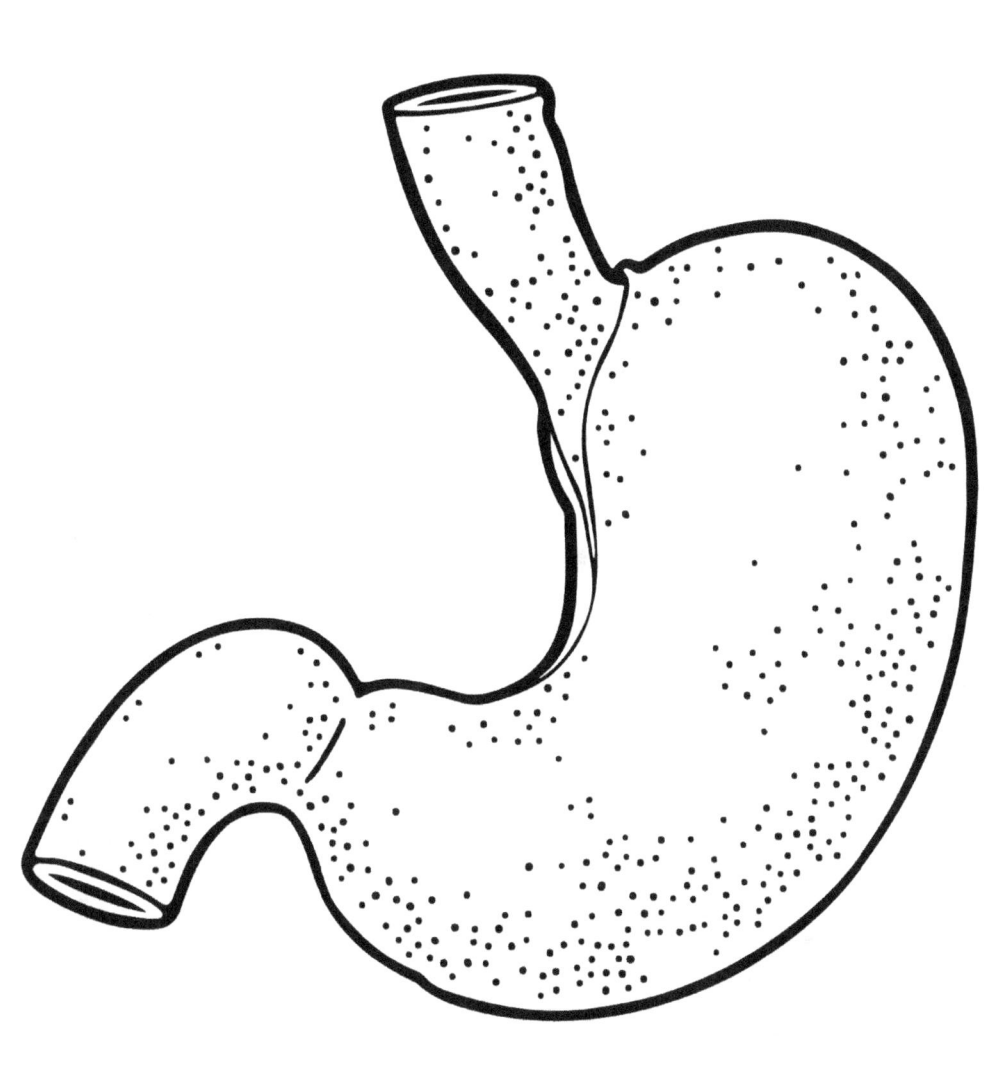

Stomach

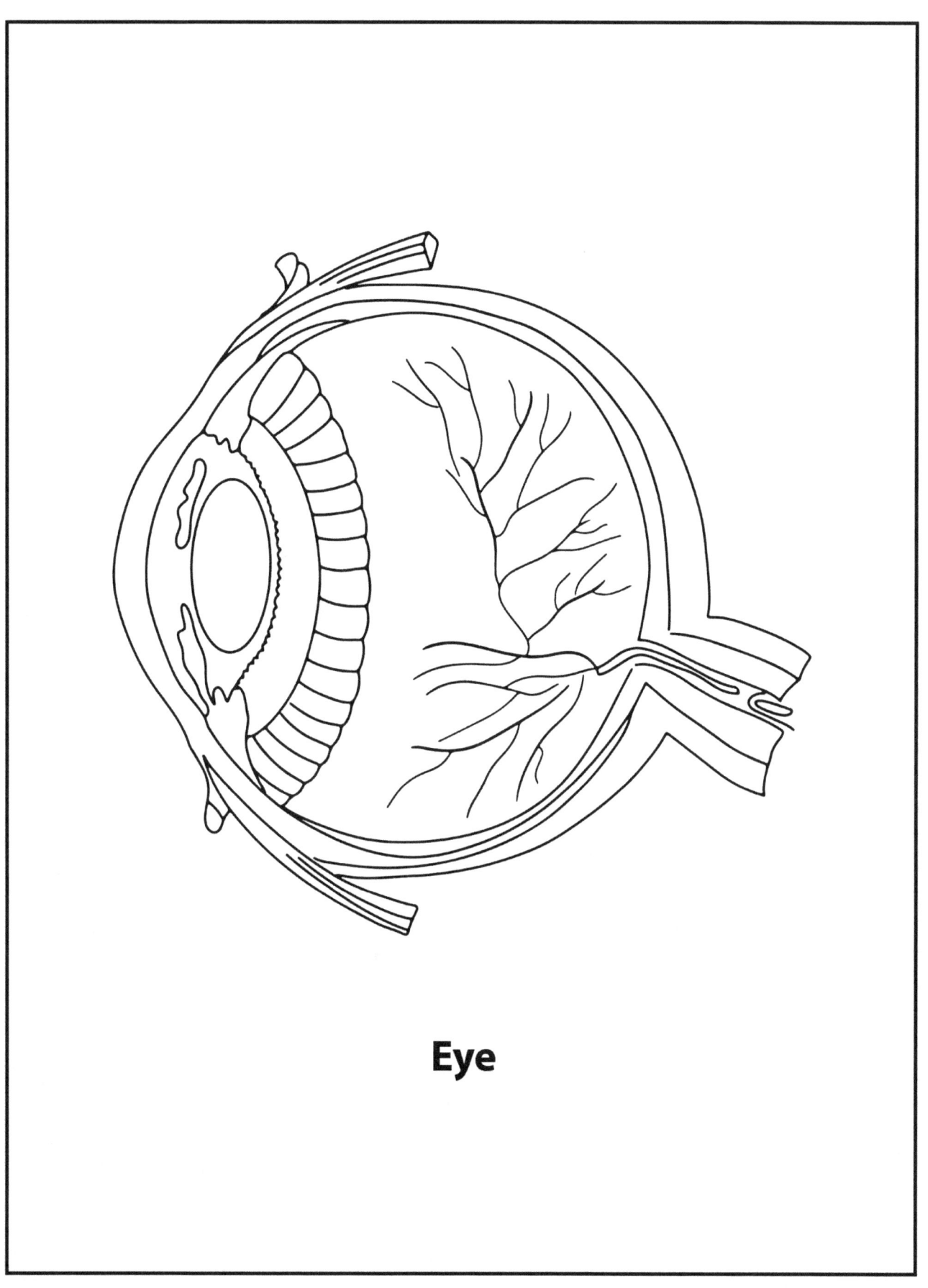

Eye

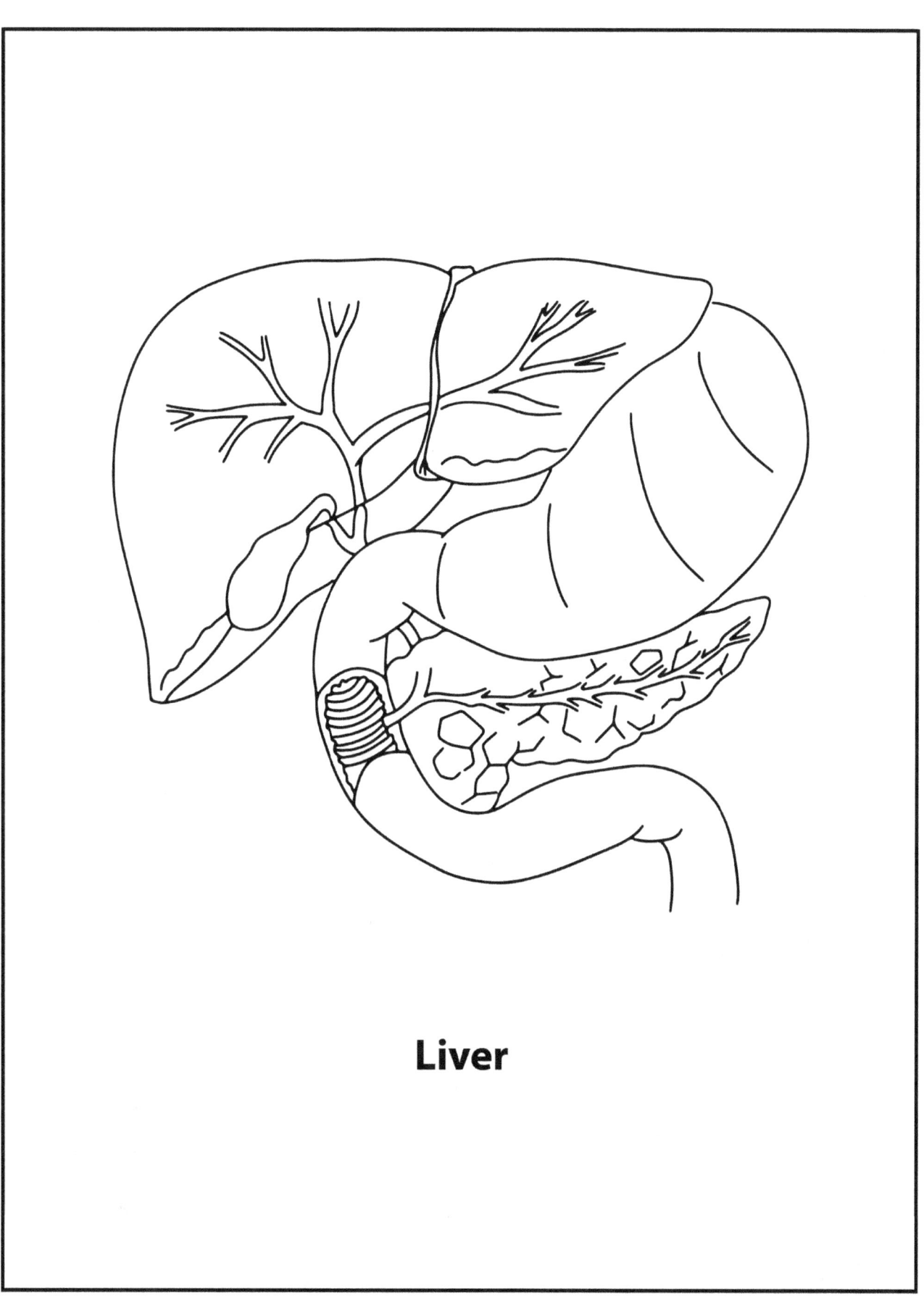

Liver

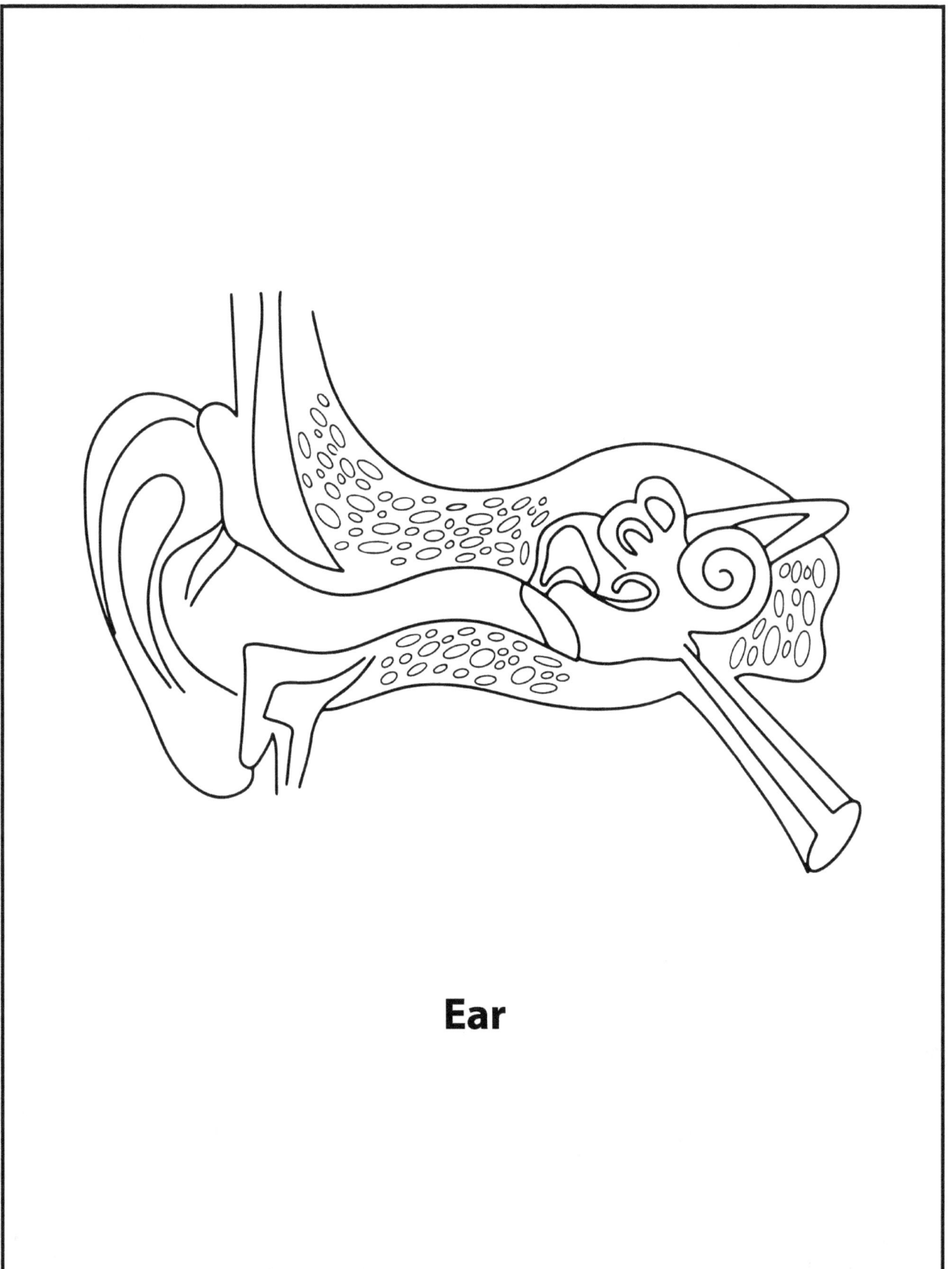

Ear

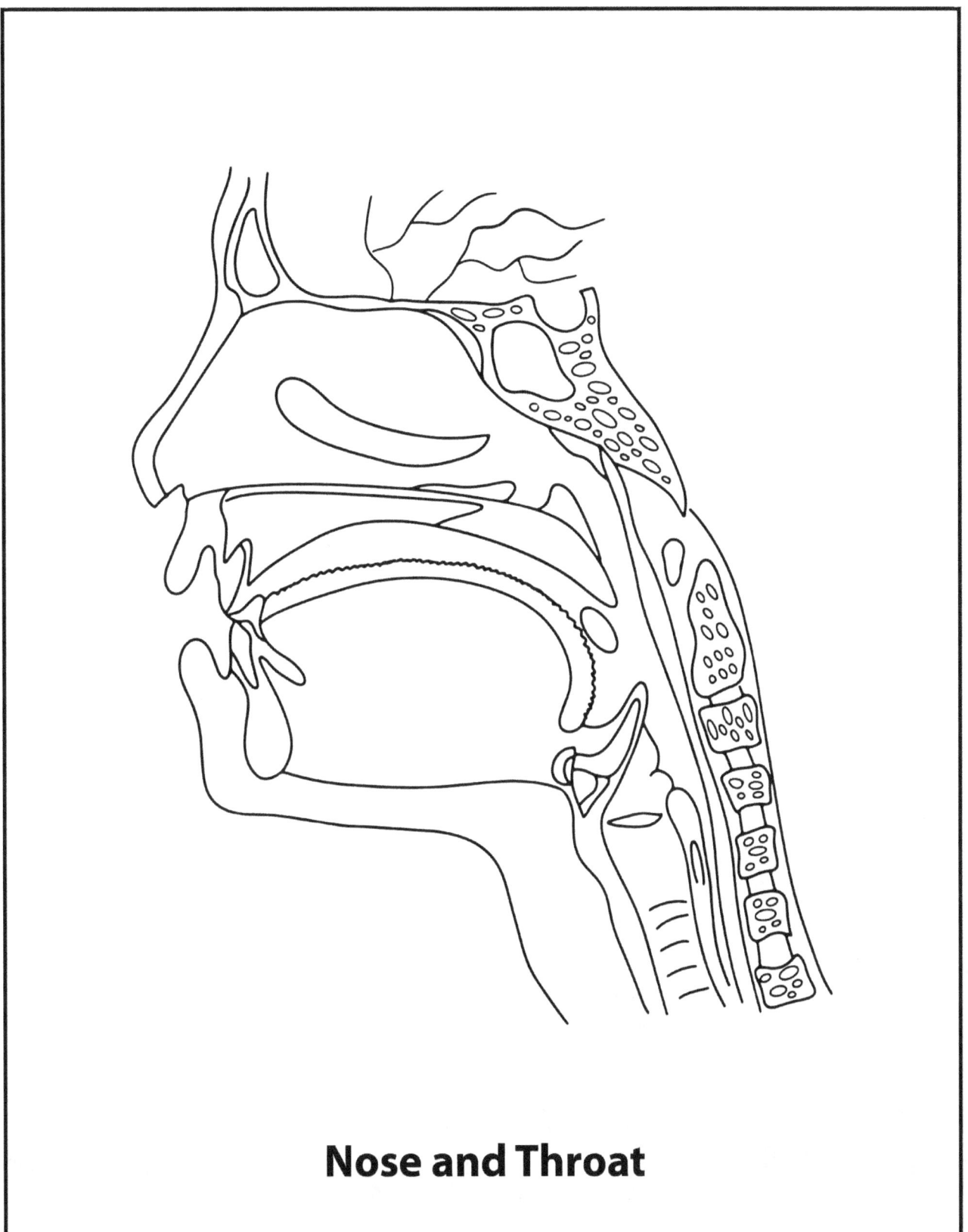

Nose and Throat

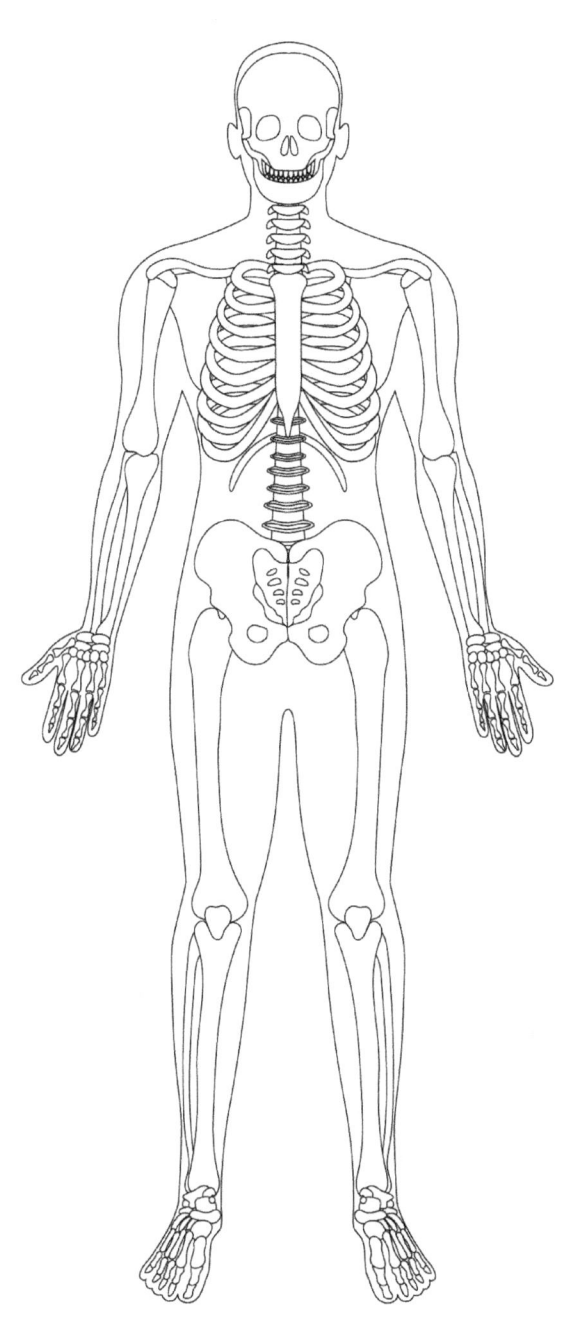

Skeletal System

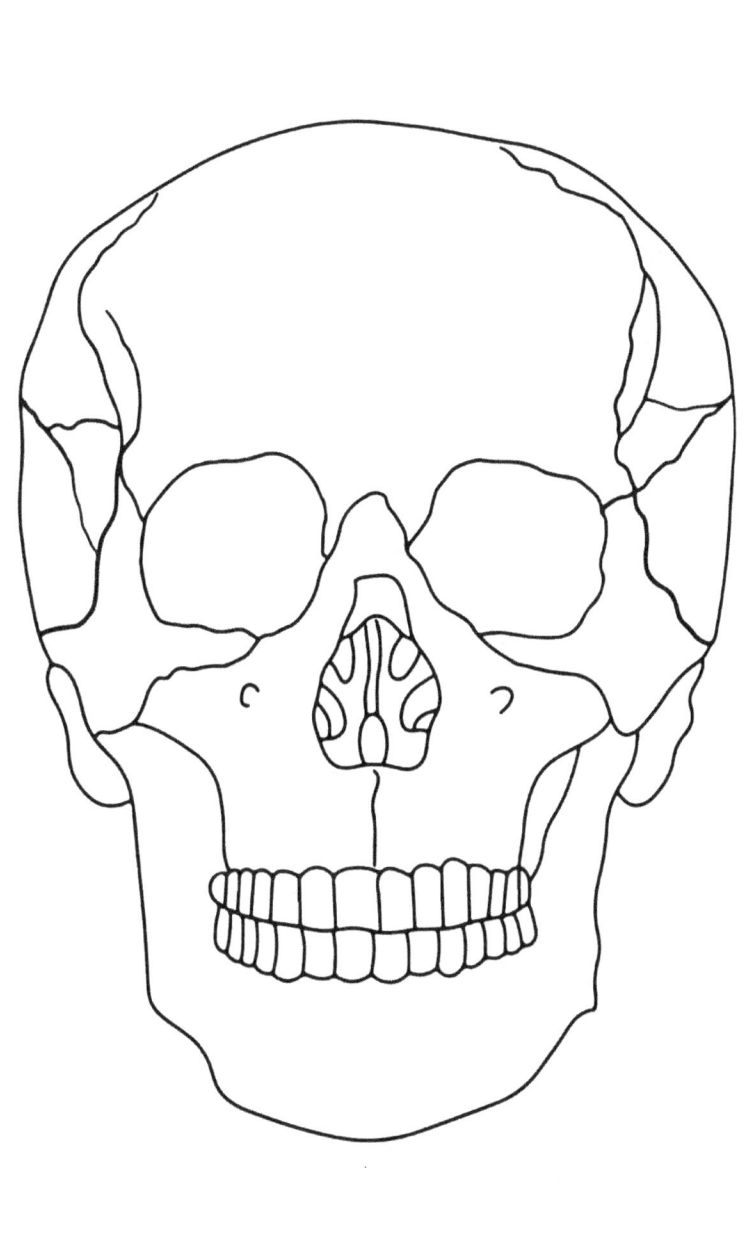

Skull

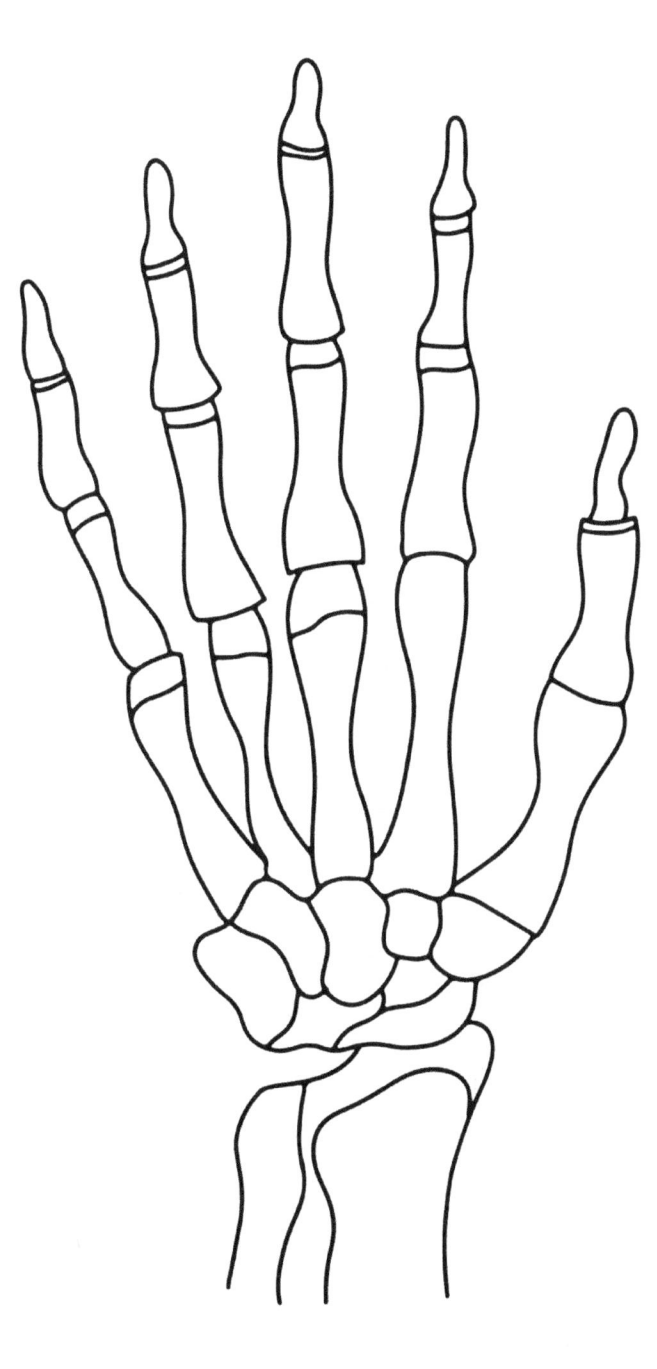

Hand and Wrist

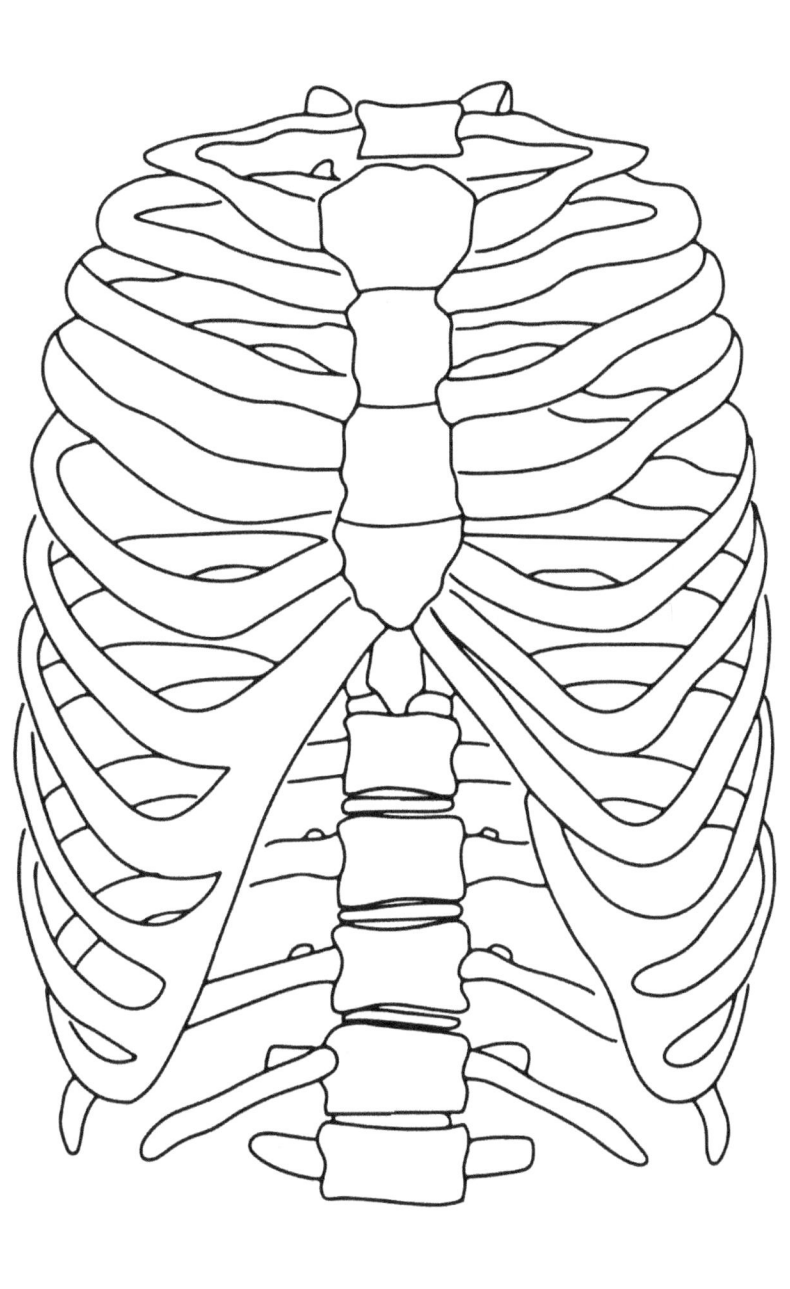

Ribs

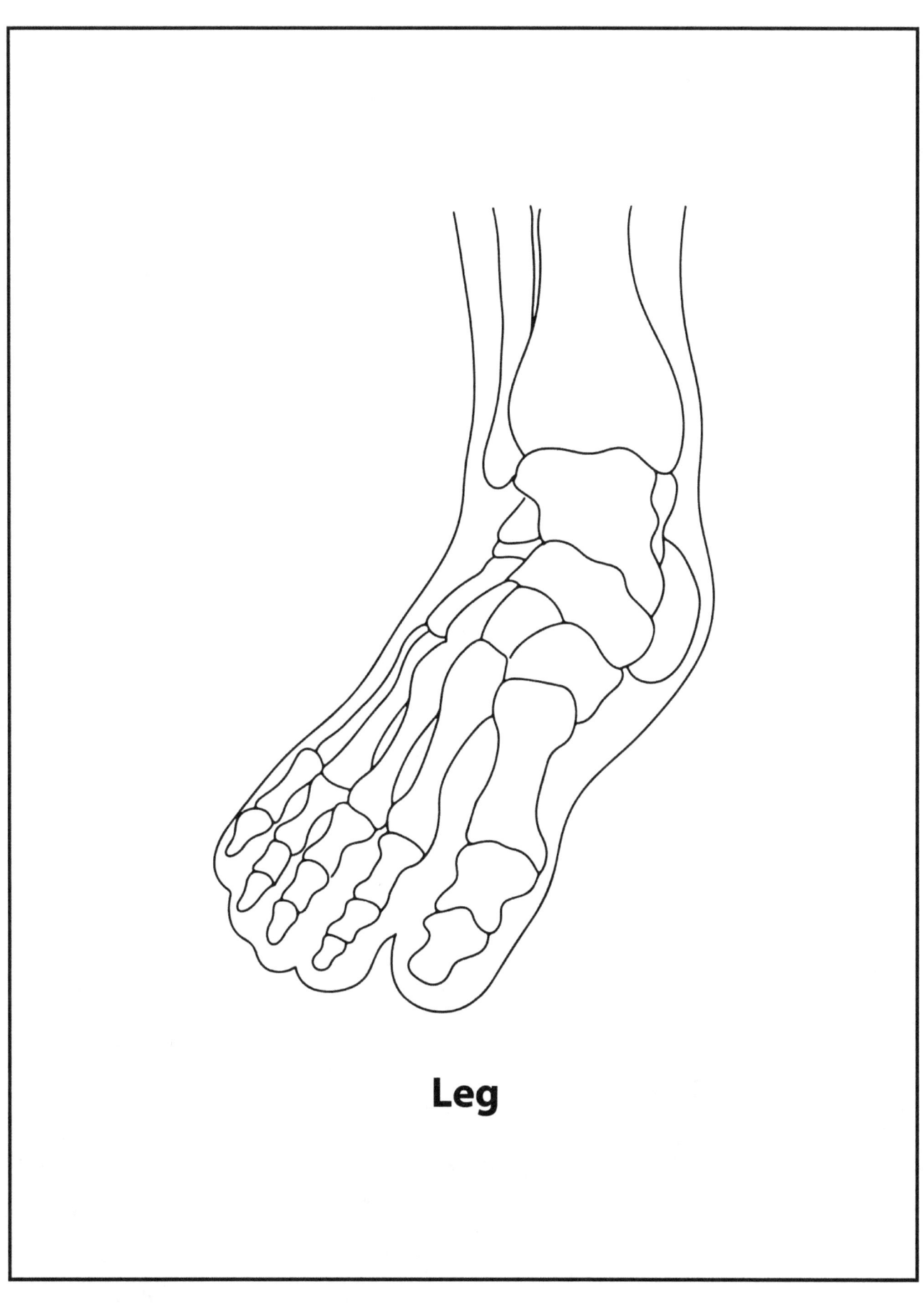

Leg

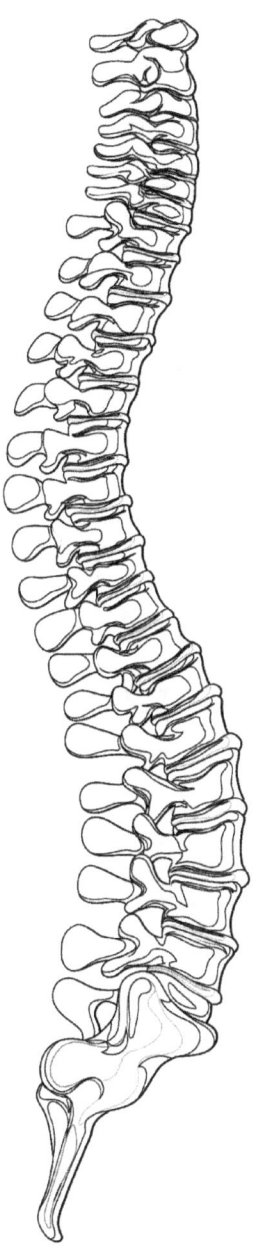

Spine

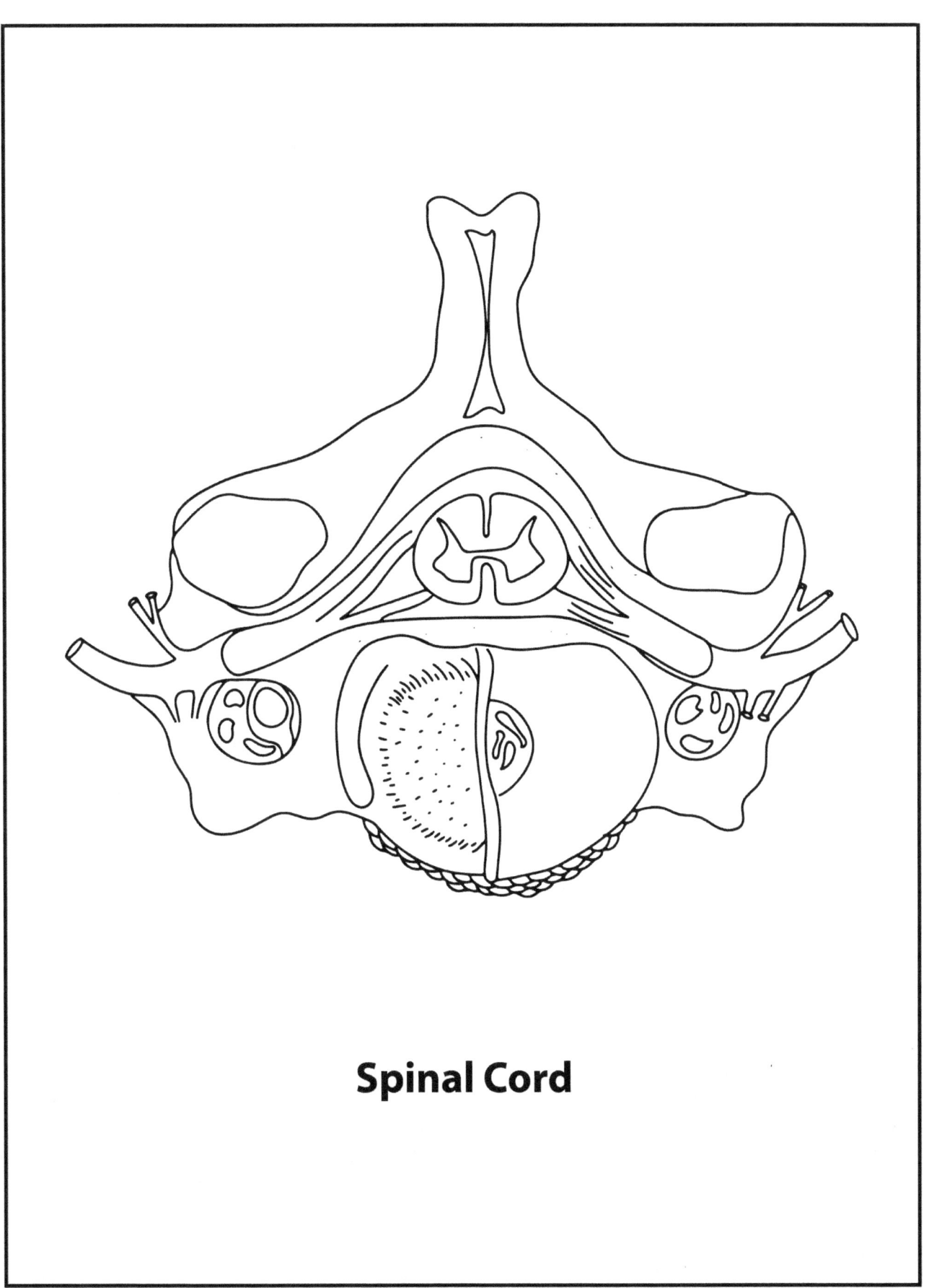

Spinal Cord

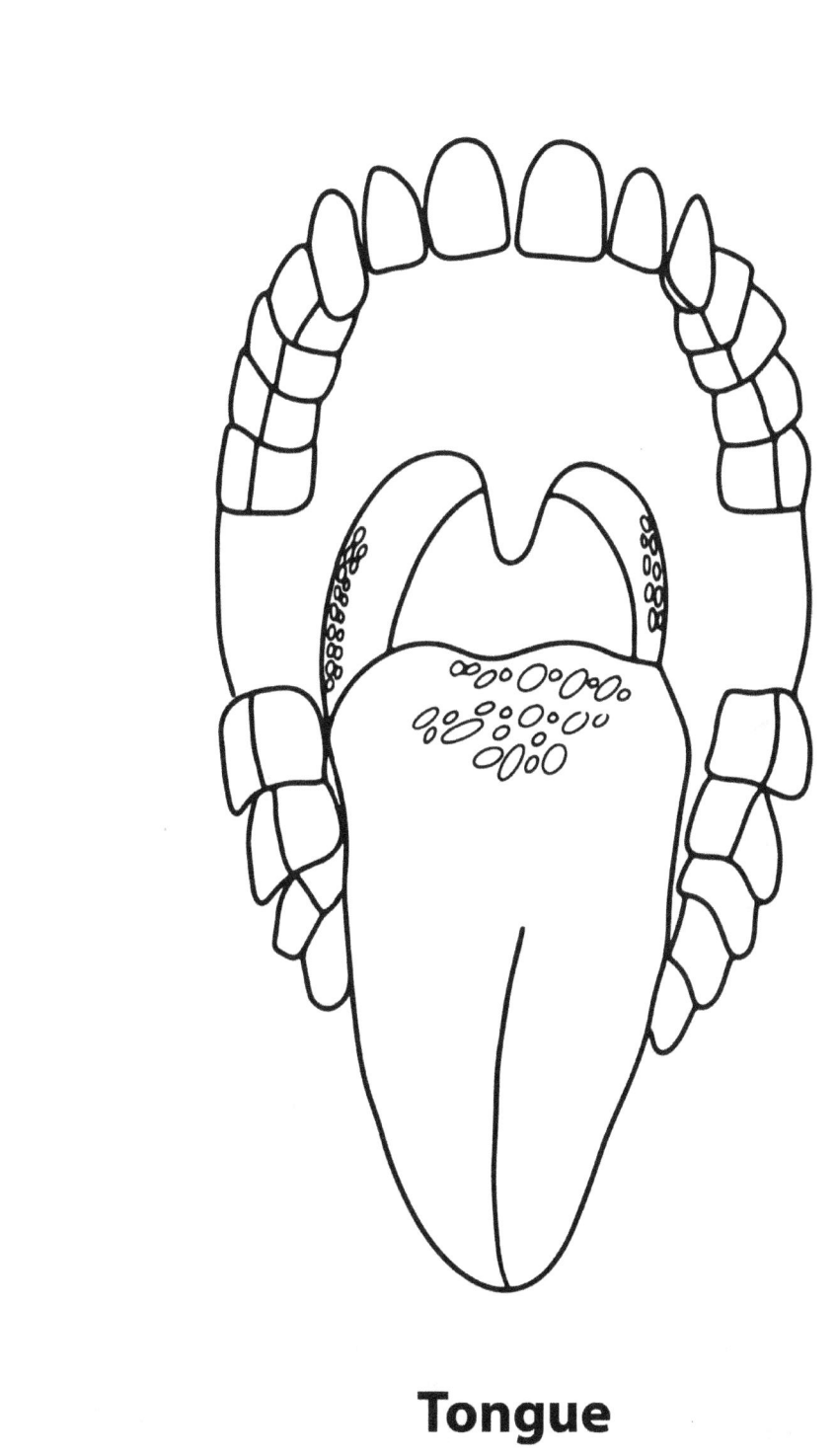

Tongue

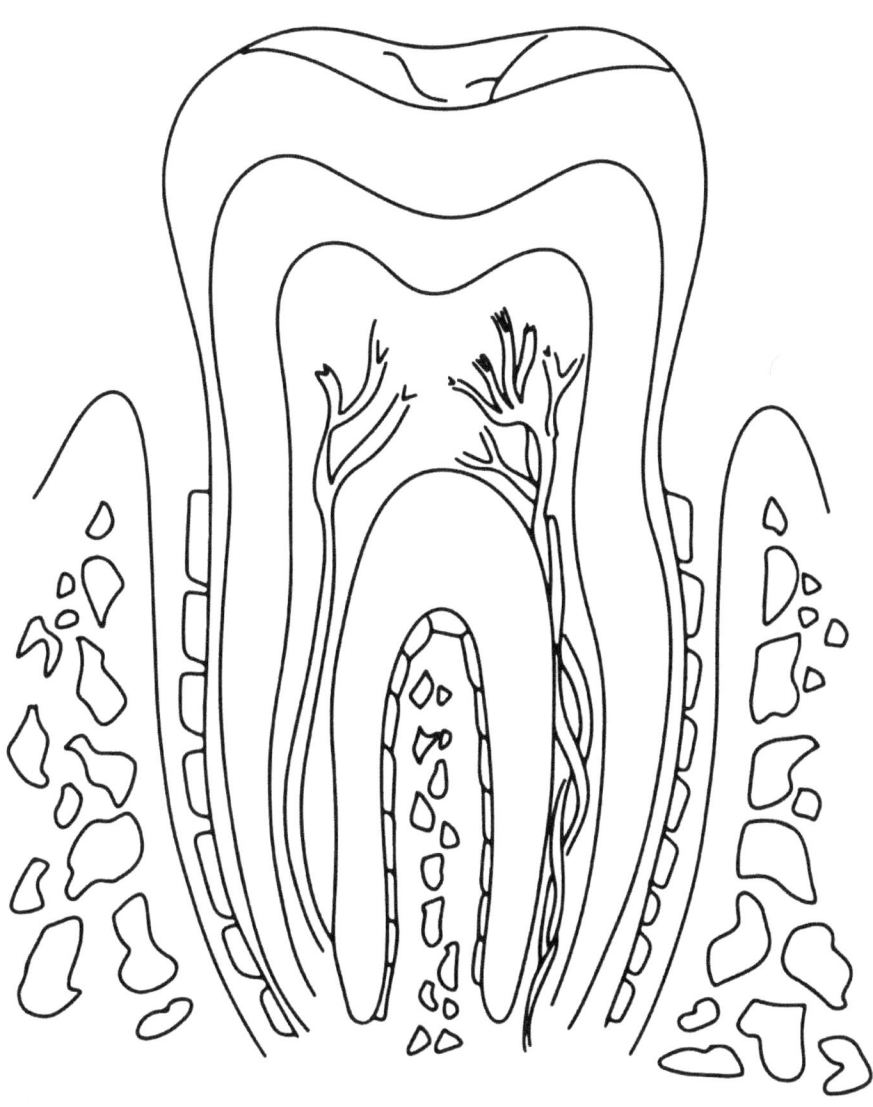

Teeth Structure

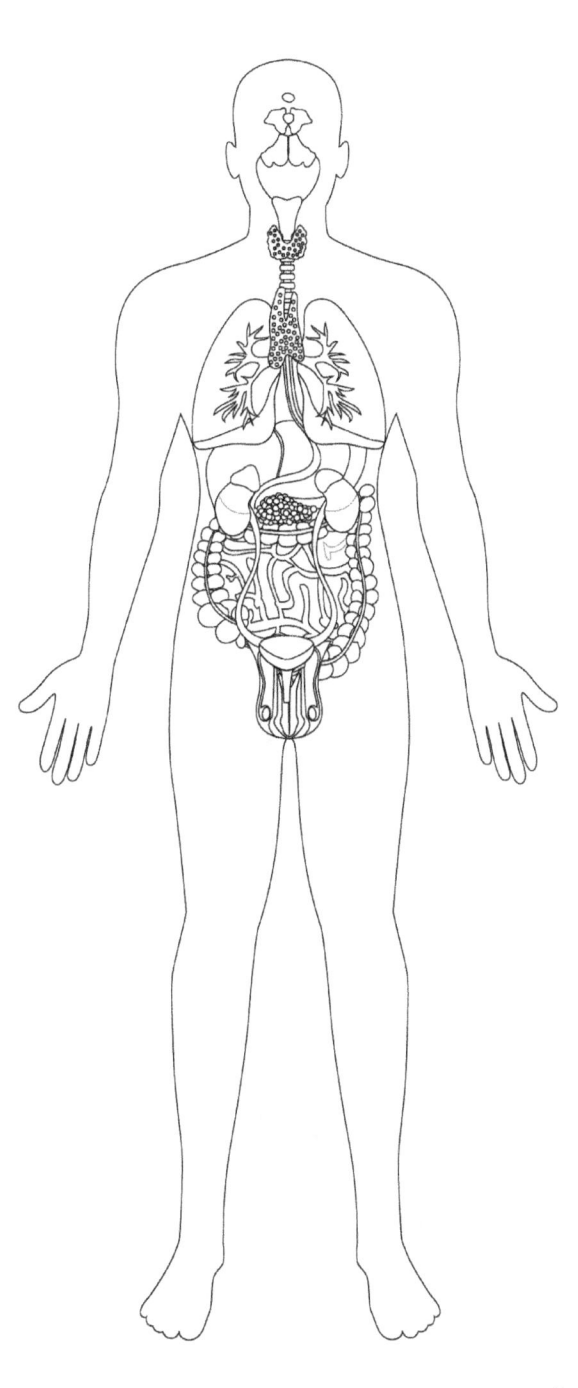

Internal Oragns

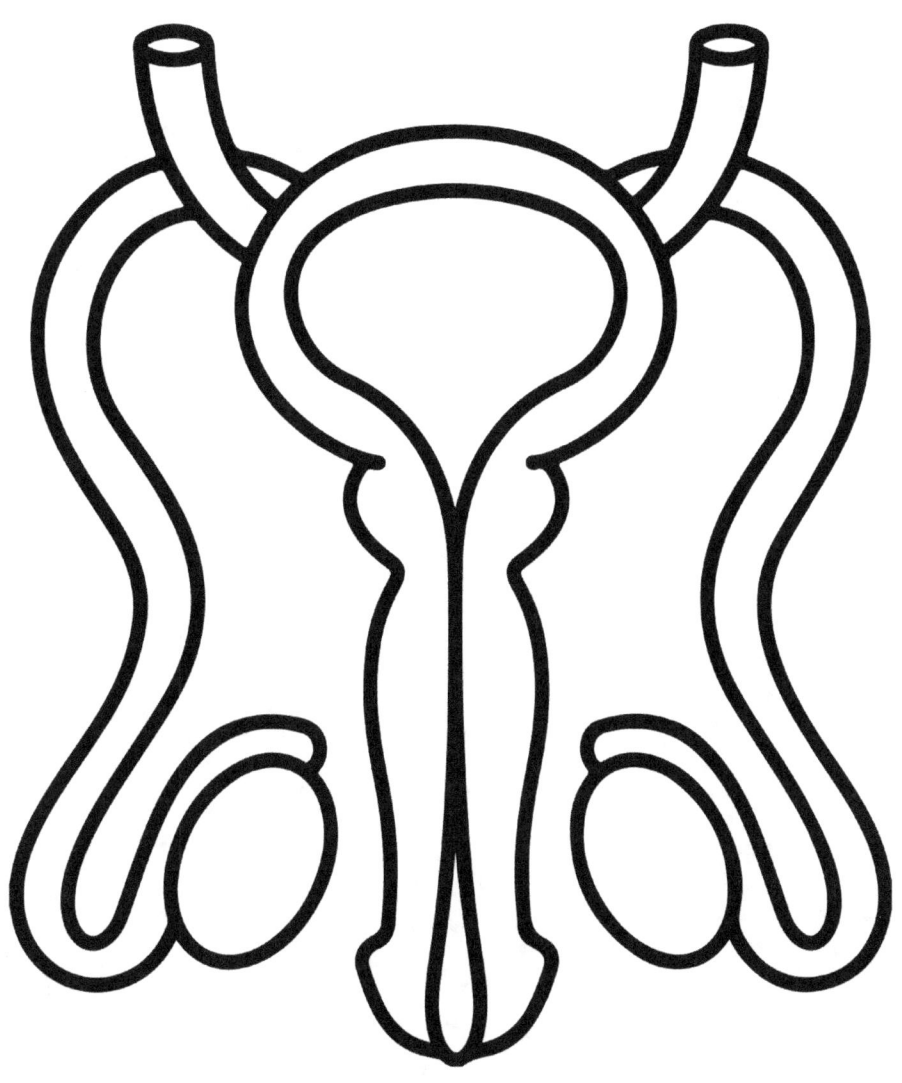

Male Reproductive System

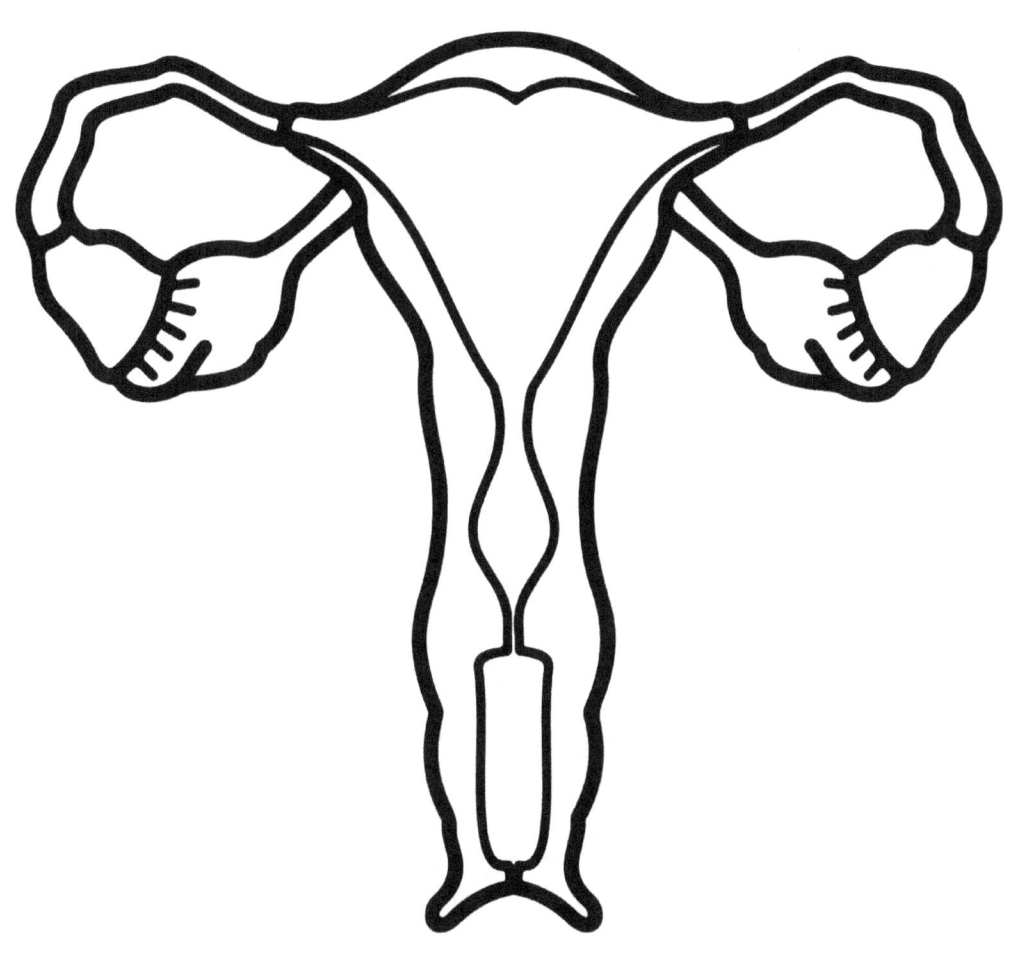

Female Reproductive System

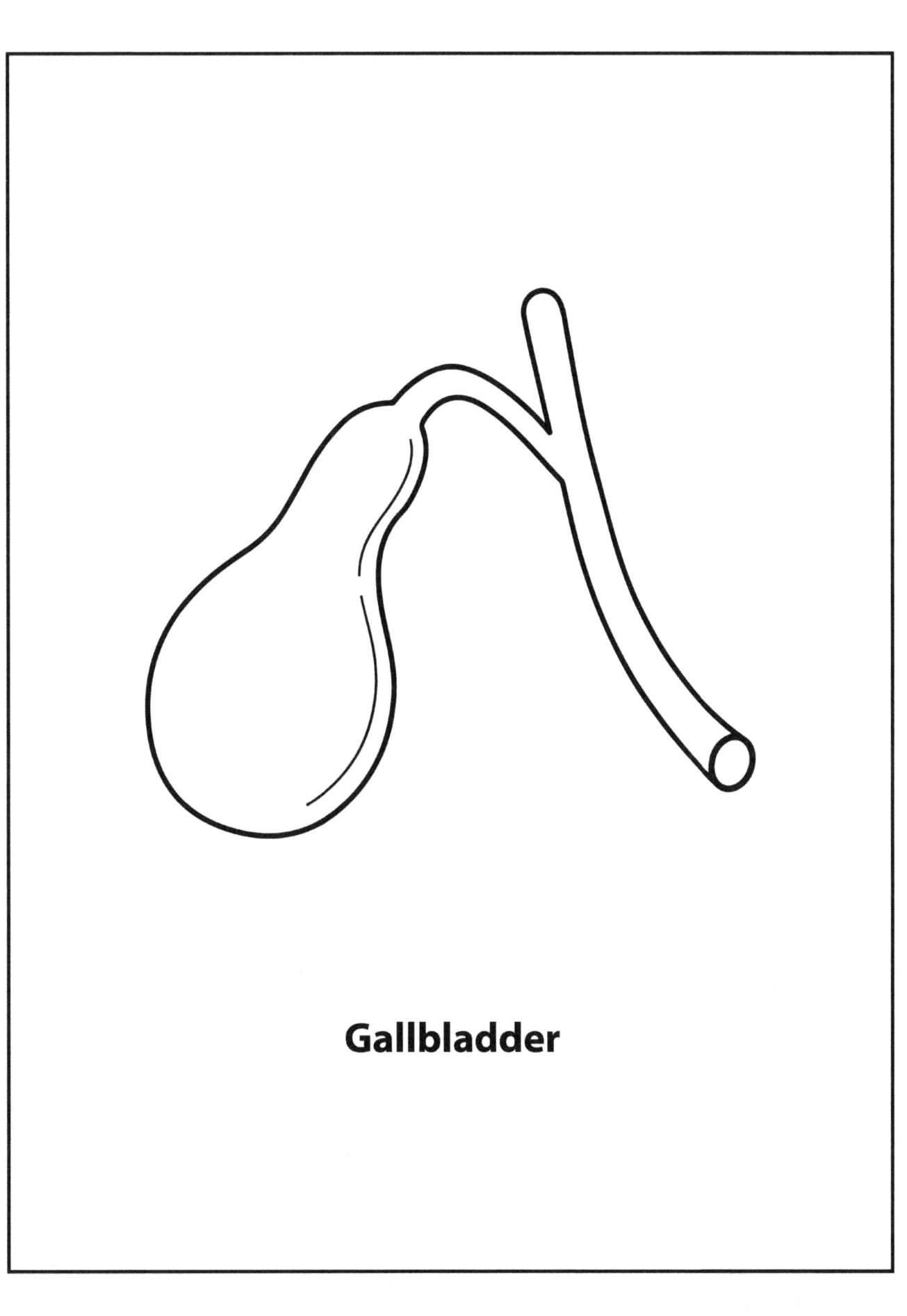

Gallbladder

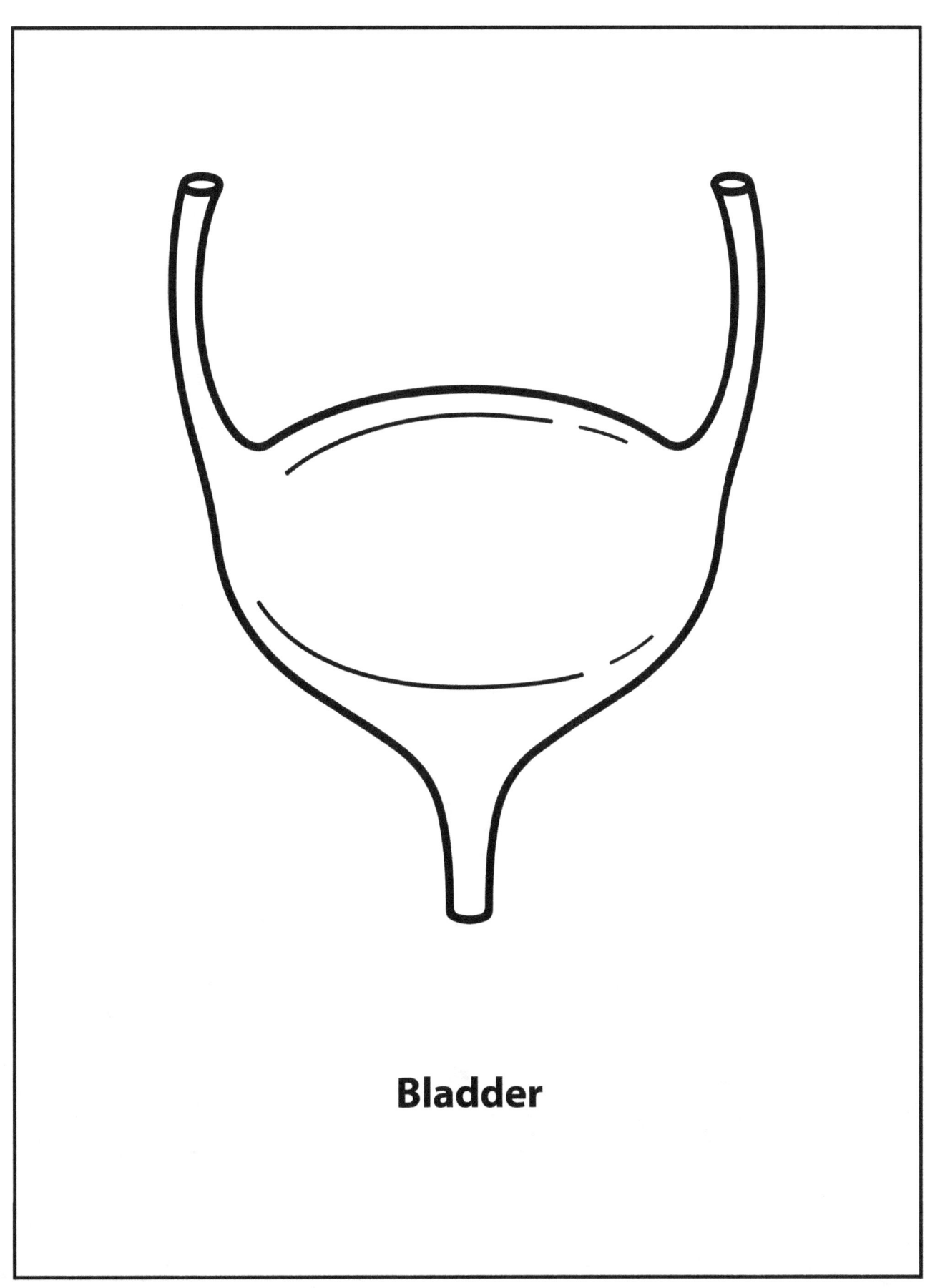

Bladder

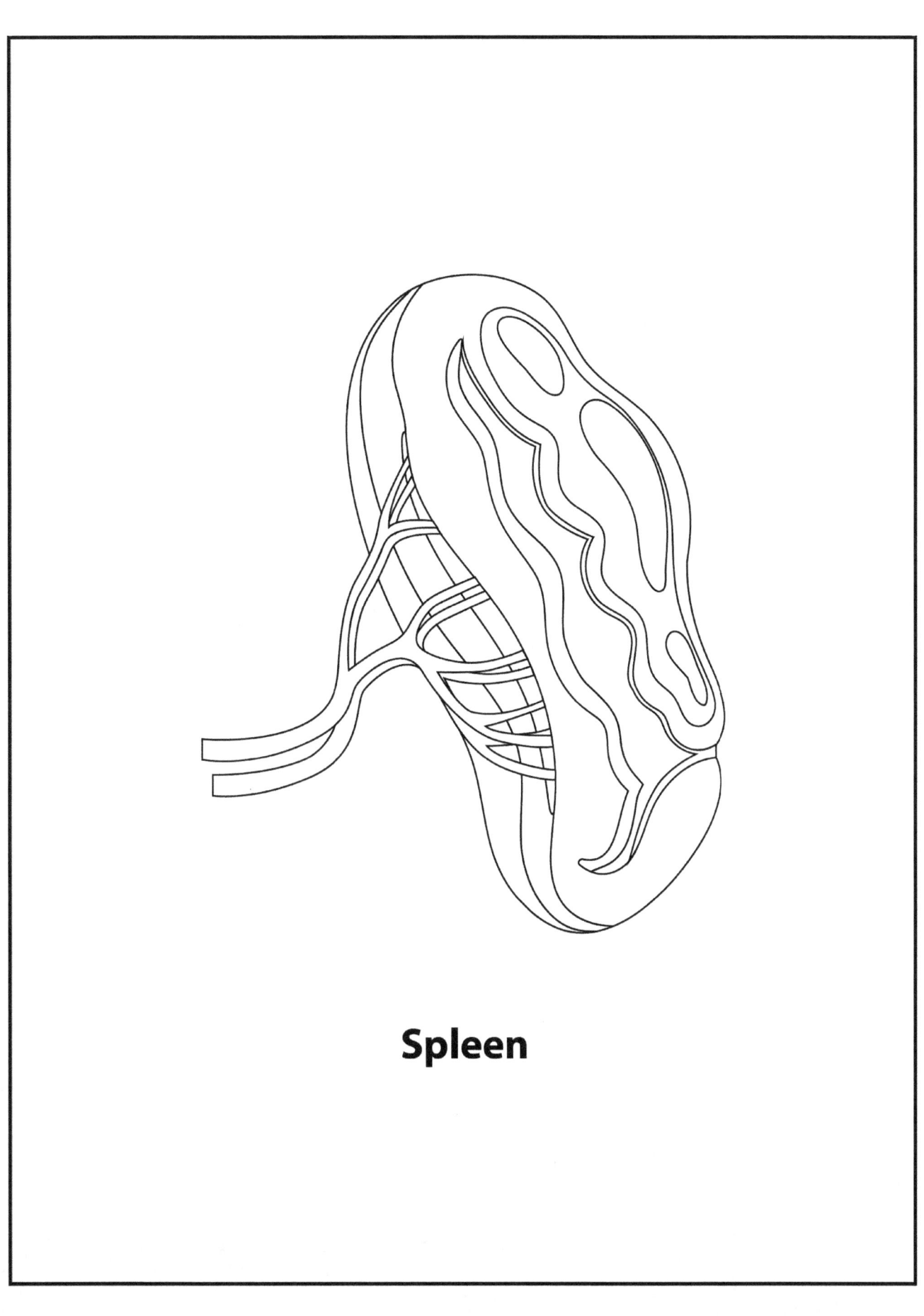

Spleen

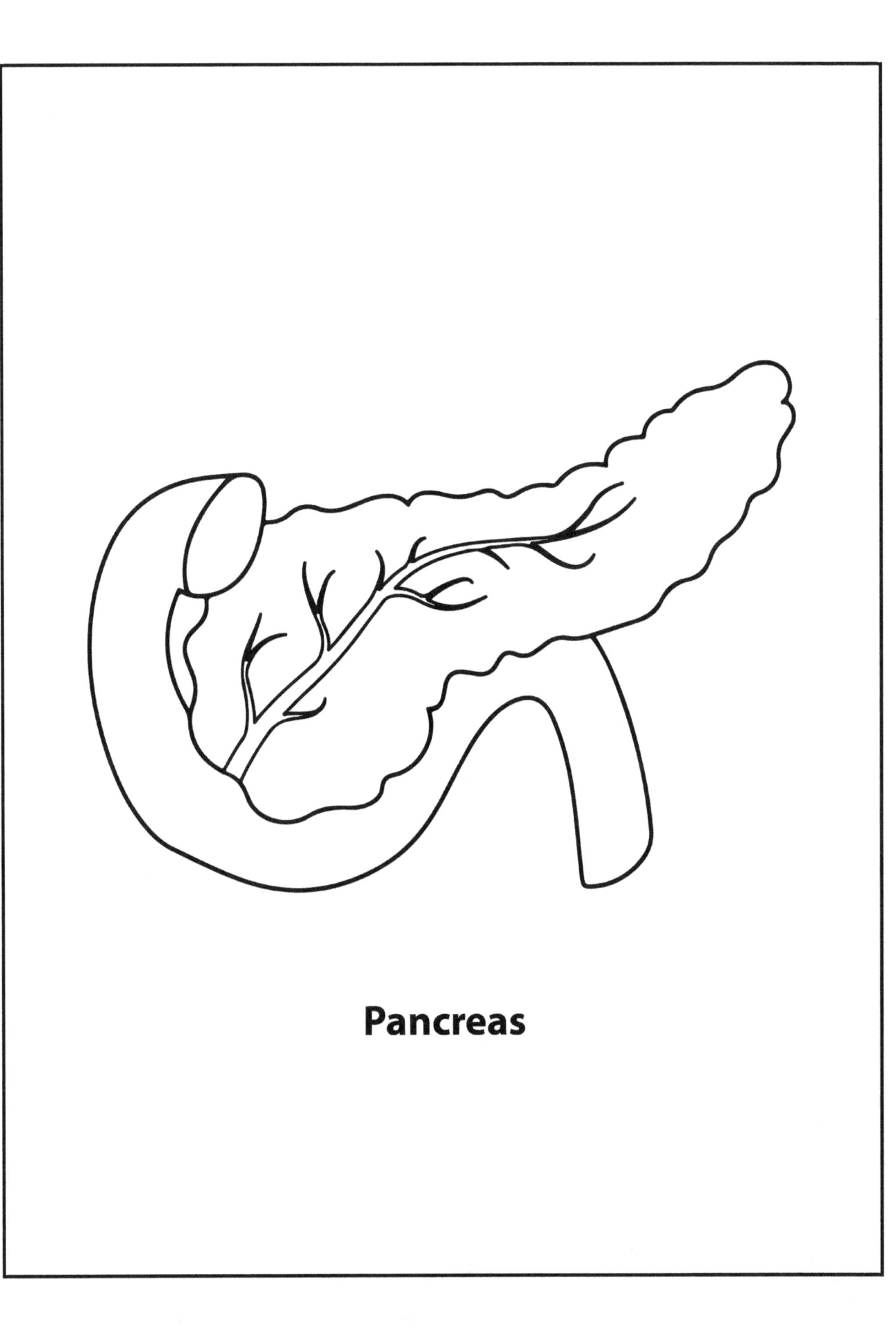

Pancreas

Skin

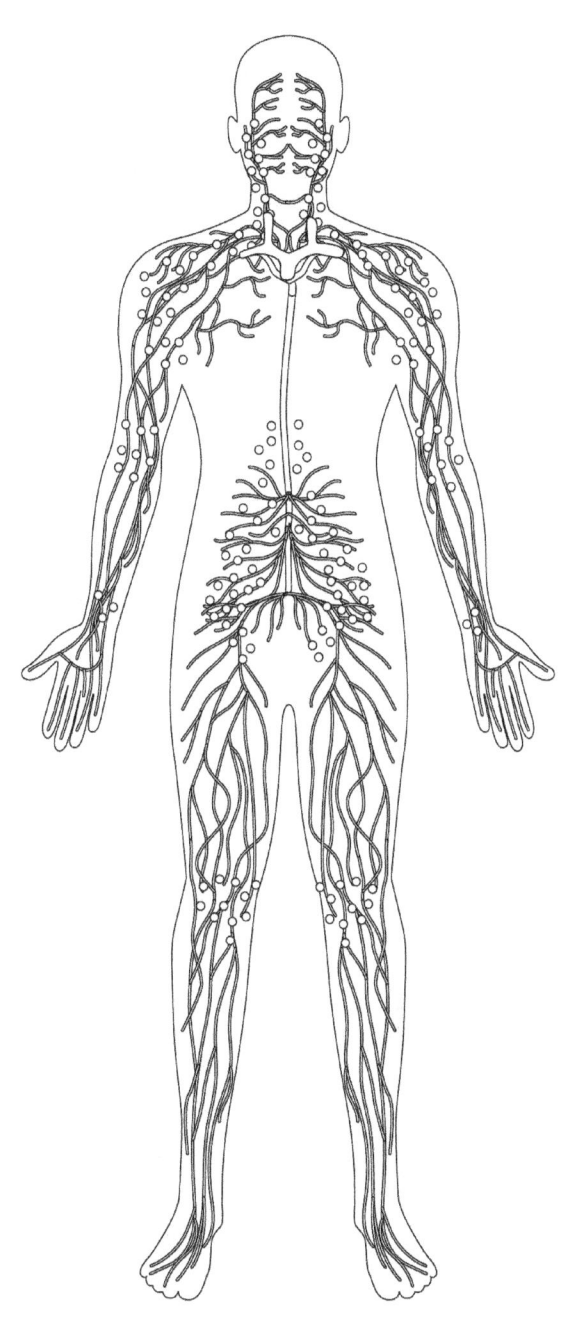

Lymphatic System

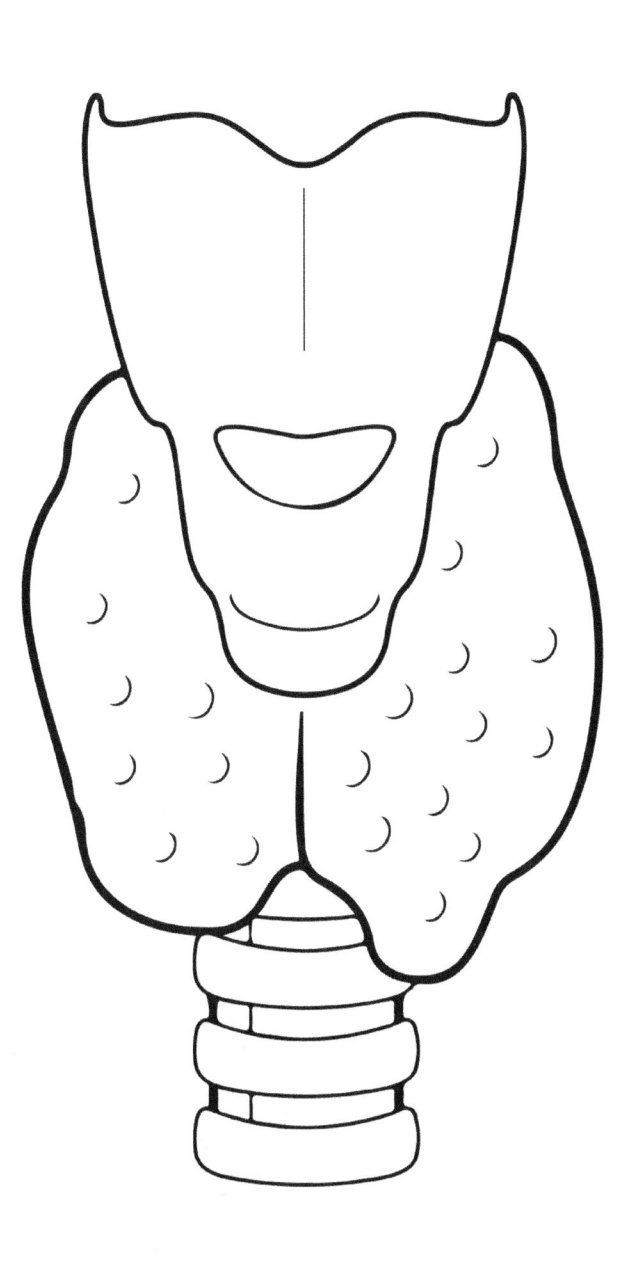

Thyroid

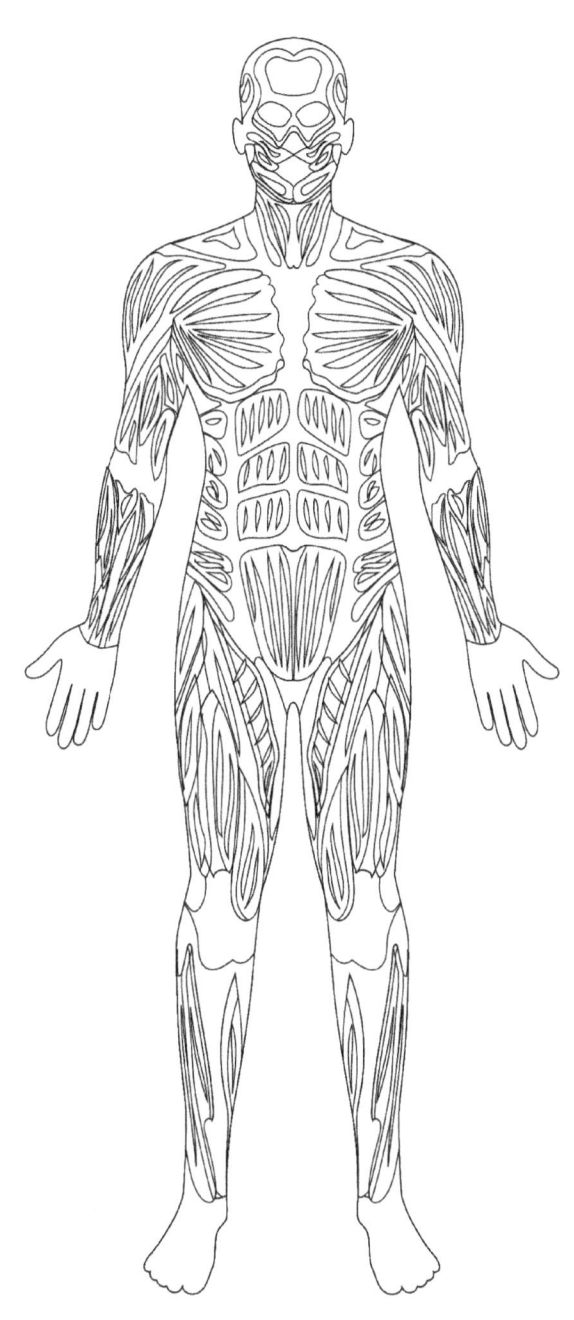

Muscle System

www.ingramcontent.com/pod-product-compliance
Lightning Source LLC
Chambersburg PA
CBHW081454220526
45466CB00008B/2647